Environmental Education

A Resource Handbook

developed by
Joe E. Heimlich

Phi Delta Kappa Educational Foundation
Bloomington, Indiana U.S.A.

Cover design by
Victoria Voelker

Cover illustration by
Christopher Ganz

Phi Delta Kappa Educational Foundation
408 North Union Street
Post Office Box 789
Bloomington, IN 47402-0789
U.S.A.

Printed in the United States of America

Library of Congress Catalog Card Number 2001098916
ISBN 0-87367-834-6
Copyright © 2002 by Joe E. Heimlich

Table of Contents

Introduction

Though environmental education is referred to as a "new" field, it has a long and colorful history. Its roots are in nature study, conservation education, and outdoor education. But environmental education truly began in the early 1960s after the publication of Rachel Carson's *Silent Spring*. It gained prominence with the growth in the awareness of environmental issues during that era.

However, the *education* part of environmental education often has been a problem. While most people tend to think of education as what happens in schools, most environmental education in the United States occurs in informal settings. Nature trails, zoos, parks, nature centers, historic and cultural sites, science centers, museums, and noncredit education programs abound with environmental education. Thus the challenge for environmental education is to establish itself in the formal education system, even though it is not a discipline or a neatly defined set of skills or beliefs.

This collection of resources was culled from the series of "EETAP Resource Library Info Sheets," which were developed for the Environmental Education and Training Partnership (EETAP) by the Ohio State University Extension and School of Natural Resources. The EETAP Resource Library was designed to ensure that educators had access to quality environmental education resources.

These articles have been categorized in six major themes in order to provide a reference guide for educators. The themes are Defining Environmental Education, What Makes Education Environmental, Integrating Environmental Education and Formal Education, Environmental Education Beyond Nature Study, Models for Infusing Environmental Education, and Environmental Education Resources.

The articles provide a host of valuable resources for educators in all disciplines and grade levels. Each article addresses one central question, concept, or idea and briefly provides the necessary background to understand the issue. Each article also lists resources available for further reading, teaching, or discussion on the topic.

The entire collection of original Info-Sheets is available online at the EETAP Resource Library repository site at www.ag.ohio-state.edu/~comm_dev/eetap.

Defining Environmental Education

Like most fields struggling for an identity, there has been a great deal of interest in defining environmental education. Through the years, it has been characterized as many things, from environmental science to environmental activism, from tree hugging to tree farming, from an elitist movement to a populist cause.

One problem in this struggle for identity has been the manner in which the popular term, "environmentalist," has been used. It has been applied to anyone who seems to take an extremist view on conservation debates, as well as to anyone who professes a nonscientific but nature-based view of an issue. In the news media, it is a label that is given to anyone who speaks on an environmental topic, regardless of their credentials. Though the term can refer to both positive and negative aspects of people engaged in environmental issues, it often is used as a derogatory label.

The challenge for environmental education is to avoid being identified with the negative connotations of the term while still maintaining the connection with environmentalists.

So just what is environmental education? The following articles provide brief summaries of the many studies, international forums, and long discussions of the field about itself. It is, after all, the ability of a field to distinguish itself from others that is the first mark of a profession.

Environmental Education: Can It Be Defined?

by Sabiha S. Daudi and Joe E. Heimlich

Environmental education has different meanings to people depending on their continuum of understanding and school of thought. This understanding may depend on their experiences, professional and social backgrounds, academic level, and learning achievements. Policy makers usually understand environmental education to be totally related to "bugs and bunnies" in the out of doors. Environmental educators consider experiential learning and behavioral change to be directly related to environmental education. Likewise, there is also a difference of opinion about the nature of environmental education among scientists. Those dealing with the natural sciences might limit themselves to conservation and preservation of ecosystems and natural resources, while social scientists consider successful environmental education to be indicative of model human behavior for management and protection of the environment. Economists, on the other hand, have such human benefits as monetary gains high on their agenda.

Given this scenario of conflicting opinions, it becomes very difficult for the learner in the field of environmental education to define it in terms that can encompass all strands. The question then is, can environmental education be defined as a single subject that focuses on the art and science of *nature*, or is it possible to bring real-life issues and the human dimensions into the arena of environmental education?

When we explore the history and development of the environmental education movement in the United States, a number of interesting facts emerge. It appears that public awareness of the state of the environment was sparked by Earth Day, celebrated on April 22 in 1970.

However, in reality the seed for environmental education has been around for a long time. Before being labeled *environmental education*, the movement had been active as nature study, conservation education (or even preservation education by some), and outdoor education. Environmental concerns of the citizens also have been addressed under the guise of science, art, geography, social studies, and citizenship education. The ultimate goal in all of these efforts has been to understand the relationship between the biotic and abiotic environment and the role human beings play in catalyzing changes in the natural world. It is the later concept that has caught the attention of many thinkers, philosophers, and educators in recent decades. This sector of the intelligencia has been prominent in giving shape to the present structure of the environmental education movement.

Whatever the popular thinking about the definition of environmental education may be, there definitely seems to be consensus on the following points:

- Environmental education is a continuous learning process that evolves according to our experiences as we go through life;
- The ultimate goal, to be achieved through experiential learning, is a change in human behavior; and
- Our educational efforts need to focus on adding a sustainable and environmentally friendly quality to life. If these are the dimensions that need to be addressed, than the educator dealing with environmental issues needs to provide positive learning experiences; encourage and acknowledge any behavioral changes, however minor, in the learners; and present sustainable ways of life practiced in different parts of the world to their learners.

To support the educators in achieving these objectives, we have identified a number of educational resources on the databases of ENC and ERIC/CSMEE that deal with definitions and experiences in environmental education.

Resources from ERIC

Heimlich, J.E. "Nonformal Environmental Education: Towards a Working Definition." *ERIC/CSMEE Information Bulletin.* Ohio, 1993. ED360154.

This information bulletin examines a taxonomy of four learning environments and explores the application of nonformal learning theory into practice in the arena of environmental education. One of the four sections establishes definitions for four learning environments.

Phipps, M.L. *Definitions of Outdoor Recreation and Other Associated Terminology.* Colorado, 1991. ED335189.

This document defines terms related to outdoor recreation, such as outdoor recreation, outdoor education, environmental education, experiential education, adventure education, wilderness education, and commercial recreation.

Miller, J. *Model Learner Outcomes for Environmental Education.* St. Paul: Minnesota State Department of Education, 1991. ED354155.

This document provides Minnesota educators with a means of systematically viewing environmental education programs within their schools and provides a tool for integrating environmental education into all courses and programs. Also included is a glossary.

Disinger, J.F. "What Research Says: Environmental Education's Definitional Problem." *School, Science, Mathematics* 85 (January 1985). EJ312591.

This article is a representative selection of definitions for the purpose of comparing and contrasting them. It also explores the origin of the term *environmental education* and defines its primary antecedents, such as *nature study*, *outdoor education*, and *conservation education*.

Resources from ENC

Cornell, J. *Sharing the Joy of Nature: Nature Activities for All Ages.* Nevada City, Calif.: Dawn, 1989. ENC-001437.

This book of activities provides young learners with experiences in the natural environment. It uses nature to stimulate joyful insights and experiences for children. The activities follow the four-stage, flow learning system, namely: awaken enthusiasm, focus attention, direct experience, and share inspiration.

Samuel, R.H. "Impediments to Implementing Environmental Education." *Journal of Environmental Education* 25 (Fall 1993). Madison, Wis.: Dembar Educational Research Services. ENC-002308.

This article focuses on the theme that implementation of a sound environmental program requires that teachers understand the philosophical and pedagogical nature of environmental education and relate it to their subject areas. The case study also reveals that planning ahead and teacher participation in decision making are imperative.

Crow, T., ed. "Earth Day in the Classroom: Mathematics and Science Materials and Resources for Teachers." *ENC Focus*, Issue 2. Columbus, Ohio: Eisenhower National Clearinghouse, 1994-1995. ENC-002301.

This guide offers teachers a sample of educational resources that can be used to highlight environmental issues in the classroom. Also included are selected readings, relevant Internet sites, and other federal projects and resources.

Evolution of Environmental Education: Historical Development

by Sabiha S. Daudi and Joe E. Heimlich

A continuing dilemma for those concerned with environmental education is the matter of definition. There are some who strive to achieve universal agreement on a precise meaning and discrete set of descriptive parameters for environmental education, but others prefer not to expend energy on what they perceive as an inherently nonproductive exercise. There also exists a third population, a potpourri of groups and individuals who have independently forwarded a variety of definitions and descriptive statements, on occasion demonstrating strong overlap and, on occasion, equally virulent disagreement.

The need and desire for a generally accepted definition has existed for at least three decades, and attempts have been made to clarify elements and boundaries of environmental education. It is useful at this point to summarize a representative selection of definitions for the purpose of comparison.

Origin of the Term

Some of the early, as well as evolving, applications of the term *environmental education* are listed in chronological order.

1948: Thomas Pritchard, Deputy Director of the Nature Conservancy in Wales, identified the need for an educational approach to the synthesis of the natural and social sciences, suggesting that it might be called *environmental education.*

1957: Brennen used the term in an article in the *Bulletin of the Massachusetts Audubon Society.*

1964: Brennen applied the term in his address to the American Association for Advancement of Science.

1979: Brennan acknowledged his early use of the term but disavowed any intention of using it other than as a synonym for conservation education.

The term *environmental education* is established, yet its meaning continues to evolve.

It is frequently acknowledged that the primary antecedents of environmental education were nature study, outdoor education, and conservation education. Nash (1976) summarized:

> The roots of environmental education lie in the same area and mentality as the beginnings of reaction against the university ideal [of compartmentalization of education - Ed.]. As early as 1891, Wilbur Jackman's *Nature Study for the Common Schools* launched a *nature study* movement which took students outdoors to explore an indivisible environment with an integrated academic approach. *Outdoor education,* as it was called by theorists such as L.B. Sharp and Julian Smith in the 1920s, had a very similar purpose. Nature study and outdoor education forced an appreciation of the multiplicity of factors that the classroom tended to isolate. Knowledge was integrated by an integrated environment. The 'Dust Bowl' mentality of the 1930s gave rise to *conservation education.* Its primary objective was to awaken Americans to environmental problems and the importance of conserving various natural resources. Because conservation education focused on problems which themselves were products of many interrelated factors, students exposed to such programs pursued a more integrated learning program.

Several other educational movements have been identified as forerunners or concurrent companions of environmental education, such as:

Resource-use education: A social studies "twin" of conservation education, focused more heavily on economics and geography than the natural sciences.

Progressive education: The original purpose of this movement related to making education more responsive to the needs of children. Among its accomplishments were "some curriculum reforms toward a more holistic approach to learning."

Resource management education has represented the professionalization of soil conservation, water management, game management, park management, urban and regional planning, landscape design, architecture, environmental engineering, metropolitan management, and so on (Shoenfeld 1971).

Population education became "a part of the environmental quality movement when it was recognized that the issues of population and environmental impacts were interwoven (Swan 1975).

A few "newer" elements, such as *energy education, biodiversity education*, and *marine-and-aquatic education*, may be viewed as responses to specific, environment-related concerns and are perceived by many to be subsets of environmental education. Using the same reasoning, it is possible to speak of environmental education as a necessary component of citizenship education and a primary consideration of global education.

Metamorphosis

Bowman, building on earlier work by Hone, has taken the position that environmental education began as an outgrowth of conservation education. Kirk approached the transition from a different perspective. He suggested that the pressures of 1960s, which were felt by the leaders in both outdoor education and conservation, were caused by an increased public awareness of the problems of air, water, noise, and landscape pollution and excess energy demands. The reaction produced a new product, philosophy, and approach that was labeled environmental education.

Definitions of environmental education began to appear, perhaps because of identified needs, such as those expressed by Helgeson et al. (1971):

> With the recent interest demonstrated throughout the nation in environmental problems, there is a band wagon effect which tempts many individuals and groups to declare that their particular interest is at the heart of environmental problems. . . . It is absolutely essential that any problem area that is to be studied seriously, be limited in scope. . . .

This debate continues, and in the opinion of Hungerford, et al. (1983):

> It is disconcerting (to say the least) for those in the implementation of environmental education goals to hear again the question: 'What is environmental education?' . . . We submit that environmental education does have a substantive structure that has evolved through the considerable efforts of many and that the framework has been documented formally in the literature. The question asked . . . has most certainly been answered. One would dare hope that this question could, at long last, be laid to rest . . . the field is quite definitely beyond the goal setting stage and into the business of implementation.

Acknowledgment: The authors thank Dr. John F. Disinger, professor emeritus, School of Natural Resources, Ohio State University, for sharing his article, "Environmental Education's Definitional Problem," published by ERIC/SMEAC as *Information Bulletin No. 2*, 1983. For citations to this information sheet, please refer to this article, which is available at ERIC Clearinghouse, located at 1929, Kinnear Road, Columbus, OH 43210-1080.

Environmental Education as Defined by the Practitioners

by Joe E. Heimlich and Sabiha S. Daudi

Environmental education carries various images for the practitioners in the field. For some, environmental education is a dimension of the environmental movement that gained momentum in the 1970s. However, it has been acknowledged that the primary antecedents of environmental education were nature study, outdoor education, and conservation education. The field of environmental education continues to evolve, and the definition of environmental education remains individually grounded. Some educators have identified several other education movements as forerunners or concurrent companions of environmental education. These include resource-use education, progressive education, and resource-management education.

Following are some of the early and evolving definitions of environmental education as described by John F. Disinger in his article, "Environmental Education's Definitional Problems," published in *Information Bulletin No. 2* of ERIC/SMEAC in 1983.

An early concise definition, which served as the basis of subsequent efforts, emerged from a graduate seminar in the Department of Resource Conservation and Planning of the University of Michigan's School of Natural Resources under the leadership of William Stapp in 1969. It declares that:

> Environmental education is aimed at producing a citizenry
> that is knowledgeable concerning the biophysical environ-
> ment and its associated problems, aware of how to solve these
> problems, and motivated to work towards their solution.

This statement was modified by R. Roth in 1970. Roth refer-
enced both biophysical and sociocultural environments and
stressed the management dimension.

> Environmental management education is the process of
> developing a citizenry that is:
>
> * knowledgeable of the interrelated biophysical and socio-
>   cultural of which [man] is a part;
> * aware of the associated environmental problems and
>   management alternatives of use in solving these prob-
>   lems; and
> * motivated to work towards the maintenance and develop-
>   ment of diverse environments that are optimum for living.

Brennan (1970) furnished the following definition of environmental education, based on earlier definitional discussions of
conservation education supplied by Brandwein and himself:

> [Environmental education is] that education which devel-
> ops in [man] a recognition of [his] interdependence with all
> of life and a recognition of [his] responsibility to maintain
> the environment in a manner fit for life and fit for living —
> an environment of beauty and bounty, in which [man] lives
> in harmony. The first part of environmental education
> involves development of understanding; the second, devel-
> opment of attitudes — a 'conservation ethics.'

In a 1975 editorial in *Journal of Environmental Education*,
Schoenfeld expressed concerns with respect to the ways environmental education was being approached by the U.S. Office of
Education. In 1980 he called for the initiation of integrated environmental management education as the proper touchstone for
environmental education.

The U.S. Office of Education, through the Environmental Quality Education Act, commonly called the Environmental Education
Act (U.S. Public Law 91-516, 1970) offered this definition:

For the purpose of this act, the term "environmental education" means the educational process dealing with [man's] relationship with [his] natural and manmade surroundings, and includes the relationship of population, conservation, transportation, technology, and urban and regional planning to the total human environment.

In Tanner's (1974) opinion:

Ultimately, we believe, environmental education must focus on the Spaceship Earth concept. It must deal with [man-man] or [man-] society relationships only as they affect, or are affected by, [man-] earth relationships. Other endeavors, worthwhile though they may be, are not environmental education. To be useful, a concept must be both inclusive and exclusive.

On the global/international scene, Recommendation 96 of the 1972 Stockholm Conference on the Human Environment called for the development of environmental education as one of the most critical elements of an all-out attack on the world's environmental crisis. A need identified at that conference (United Nations 1972) was:

Creating citizenries not merely aware of the crisis of overpopulation, mismanagement of natural resources, pollution, and degradation of the quality of human life, but also able to focus intelligently on the means of coping with them.

The Belgrade Charter highlighted this goal statement at the October 1975 International Environmental Education Workshop:

The goal of environmental education is to develop a world population that is aware of, and concerned about, the environment and its associated problems and which has the knowledge, skills, attitudes, motivations, and commitment to work individually and collectively toward solutions of current problems and the prevention of new ones.

The basic aim of environmental education as defined by the participants of the 1977 UNESCO-UNEP Intergovernmental Conference on Environmental Education (held in Tbilisi, Georgia, USSR), also known as the Tbilisi Declaration, is:

to succeed in making individuals and communities understand the complex nature of the natural and built environments resulting from the interactions of their physical, biological, social, economic, and cultural aspects, and acquire the knowledge, values, attitudes, and practical skills to participate in a responsible and effective way in anticipating and solving environmental problems, and in the management of the quality of the environment.

At the same time, in most of the world, the goal of survival and human welfare is considered superordinate. This causes environmental education to be "different" from other forms of education, because it implies efforts to change behavior. As Schmeider (1977) wrote:

Environmental education calls for participation in real world activities and for modification and changes — sometimes radical ones — in the attitudes and behavior of people, yet neither approach is very central to education the way it is commonly practiced throughout the world.

The above statements clearly indicate that efforts to "define" environmental education continue. At the same time, it also appears that groundwork has been laid for dialogue, early agreement, and further inquest.

If you are interested in conducting further research into the opinions of other practitioners of environmental education, please refer to resources found in the Educational Resources Information Center (ERIC) and Eisenhower National Clearinghouse (ENC) collections, and search the ERIC or ENC collections online or at a local library or university.

Acknowledgment: Thanks to John F. Disinger, professor emeritus, School of Natural Resources, Ohio State University, for sharing his article, "Environmental Education's Definitional Problem," published by ERIC/SMEAC as *Information Bulletin No. 2,* 1983.

Education for Sustainability:
Common Grounds for
Achieving Goals
of Environmental Education

by Sabiha S. Daudi and Joe E. Heimlich

Understanding global environmental issues and taking action to confront them are challenges that need to be addressed not only by educators but also by planners, economists, policy makers, natural and social scientists, and the general public. A multipronged approach to problem solving is required to deal with the complex issues and to place them in a frame of "sustainability."

The concept of *global education* deals with multiple education disciplines and the widespread concerns of educators. The issues in global education are related to consumptive lifestyles, poverty, increasing population, diminishing non-renewable energy sources, and management of available natural resources. The overarching concern for educators is finding and promoting sustainable ways of living on this planet.

According to the President's Council on Sustainable Development (1997):

> Education for Sustainability is the continual refinement
> of the knowledge and skills that lead to an informed citizenry
> that is committed to responsible individual and collaborative

actions that will result in ecologically sound, economically prosperous, and equitable society for present and future generations. The principles underlying education for sustainability include, but are not limited to, strong core academics, understanding the relationships between disciplines, systems thinking, lifelong learning, hands-on experiential learning, community-based learning, technology, partnerships, family involvement, and personal responsibility.

On the global level, *Agenda 21*, the document produced by the 1992 United Nations Conference on the Environment and Development, declares that education is "critical for promoting sustainable development." Education for sustainability, in which many other disciplines are indispensable components, will engage partners from all arenas — adult education, on-the-job training, other formal and nonformal education programs, and the media — to reach out to as many individuals as possible. This was a recommendation developed in a demonstration project of the President's Council on Sustainable Development, titled "National Forum on Partnerships Supporting Education about the Environment." It was held at the Presidio in San Francisco in the fall of 1994, and the results were published in a report, *Education for Sustainability: An Agenda for Action.*

As environmental educators, we need to understand the comprehensive nature of education for sustainability — a concept that encompasses not only environmental education, but also education related to population, health, resource management, and civic responsibilities of individuals. There are a number of teaching and learning resources available to address this.

Resources from ERIC

Merryfield, M., *Teacher Education in Global & International Education*. Washington, D.C.: American Association of Colleges for Teacher Education, 1994. ED376166.

Global and international education must be a high priority of higher education and teacher education if students are to develop the knowledge, skills, and attitudes that are necessary for decision-making and effective participation in a world characterized by interconnectedness, cultural pluralism, and increasing competition for resources.

Tye, K.A., ed. *Global Education: From Thought to Action. The 1991 ASCD Yearbook.* Alexandria, Va.: Association for Supervision and Curriculum Development, 1990. ED326970.

Viewed as a social movement for change, the global education movement calls for the infusion of a global perspective into all curriculum areas. This yearbook defines global education, explains its importance, describes its implementation, and demonstrates its uses for school improvement.

Merryfield, M.M., et al., eds. *Preparing Teachers to Teach Global Perspectives: A Handbook for Teacher Educators.* Thousand Oaks, Calif.: Corwin, 1997. ED404328.

This book provides a conceptual framework that encourages exploration of global perspectives. It provides teacher educators with a guide for establishing goals, objectives, rationale, and a working definition for global education.

Corson, W.H., ed. *Citizen's Guide to Sustainable Development.* Washington, D.C.: Global Tomorrow Coalition, 1993. ED383524.

This book is intended as a working guide for the citizen interested in understanding global environmental issues and taking action to confront them. Fourteen "issue" chapters document major changes resulting from the rapid growth of human numbers and their effect on Earth's resources.

Peters, R.O. *Environs: Living in Natural and Social Worlds.* Plaistow, N.H.: Global Horizons, 1993. ED372014.

The basic premise underlying this book is that in order to live and prosper in both natural and social environments, humans need to understand their origins, composition, characteristics, and life-sustaining processes. Education is a way to increase understanding of those worlds.

Resources from ENC

Exploring Sustainable Communities: Comprehensive Course Work on the Global Environment. Produced by the World Resources Institute Environmental Education Project, 1997. ENC-011146.

This teacher's guide, developed for secondary grades as part of the *Teacher's Guide World Resources* series, contains seven lessons that help students examine global trends in urbanization.

President's Council on Sustainable Development. *Education for Sustainability: An Agenda for Action*. Washington, D.C.: U.S. Government Printing Office, 1994. ENC-011710.

This government document, developed by the President's Council on Sustainable Development, describes an action plan to integrate education for sustainability into broader curricula. The PCSD, created in 1993, brought together leaders from industry, government, and environmental, labor, and civil rights organizations to develop policy recommendations.

Disinger, J.F. *Environmental Education for a Sustainable Future*. Columbus, Ohio: ERIC Clearinghouse for Science, Mathematics, and Environmental Education, 1990. ENC-011726.

This article, developed for educators by the Educational Resources Information Clearinghouse (ERIC), is a digest that summarizes the concept of sustainable resource management and discusses material available for teaching about sustainability.

Snyder, S.A., and Paden, Mary, eds. *Sustainable Development*. Washington, D.C.: World Resources Institute, 1994. ENC-008331.

This teacher's guide, developed for grades nine through 12, is part of a series of units that integrate the global environment with science, mathematics, geography, history, and civics. Each unit introduces an issue and features practical lesson plans, teaching objectives, student handouts, discussion activities, suggestions for further readings, and audiovisual resources.

Agenda 21:
Global Partnership for
Sustainable Development

by Sabiha S. Daudi and Joe E. Heimlich

Humanity stands at a defining moment in history. We are confronted with a perpetuation of disparities between and within nations; a worsening of poverty, hunger, ill health, and illiteracy; and the continuing deterioration of the ecosystems on which we depend for our well-being. However, integration of environment and development concerns, and greater attention to them, can lead to the fulfillment of basic needs, improved living standards for all, better protected and managed ecosystems, and a safer, more prosperous future. No nation can achieve this on its own; but together it is possible — in a global partnership for sustainable development.

On 14 June 1992 in Rio de Janeiro, in order to meet the challenges of environment and development, states of the world decided to establish a new partnership and adopted a global agenda. This partnership commits all signatories to engage in continuous and constructive dialogue, inspired by the need to achieve a more efficient and equitable world economy, while considering the increasing interdependence among the community of nations. A major goal of this effort is to have sustainable development become a priority item on the agenda of the international community.

This global partnership is built on the premises of General Assembly resolution 44/228 of 22 December 1989, which was adopted when the nations of the world called for the United Nations Conference on Environment and Development, and on the acceptance of the need to take a balanced and integrated approach to environment and development questions.

Agenda 21 addresses the pressing problems of the present and aims at preparing the world for the challenges of the next century. It reflects a global consensus and political commitment at the highest level on development and environment cooperation. Its successful implementation is first and foremost the responsibility of governments. National strategies, plans, policies and processes are crucial in achieving this. International cooperation provides support and supplements national efforts. In this context, the United Nations system has a key role to play. Other international, regional, and subregional organizations also are called on to contribute to this effort. The broadest public participation and the active involvement of the nongovernmental organizations and other groups also should be encouraged.

The program areas that constitute Agenda 21 are described in terms of the basis for action, objectives, activities, and means of implementation. Agenda 21 is a dynamic program. It is being implemented by the various countries according to their different situations, capacities, and priorities and in full respect of all the principles contained in the Rio Declaration on Environment and Development. It has evolved over time in the light of changing needs and circumstances. This process is a benchmark for the beginning of a new global partnership for sustainable development.

As environmental educators, we need to effectively communicate the need for cooperative efforts, be it at global, regional, or local levels, to our learners. Such partnerships could bring the goals of environmental education closer.

Resources from ERIC

Bureau for the Coordination of Environmental Programs. *Agenda 21 UNCED Follow Up.* Paris: United Nations Educational, Scientific, and Cultural Organization, 1993. ED385427.

The United Nations Conference on Environment and Development (UNCED) took place in Rio de Janeiro in 1992. The main results of UNCED were the Rio Declaration, Agenda 21, Convention on Biological Diversity, Framework Convention on Climate Change, and Statement of Forest Principles.

Boyd, D. "UNEP After Rio." *Our Planet* 4, no. 4 (1992): 8-11. EJ458253.

Government and United Nations officials, environmentalists, reporters, and others share their perceptions of the results and significance of the Earth Summit for the planet, governments, the United Nations, organizations, and themselves as individuals. This article also discusses Agenda 21 and its implication strategies.

Harvey, T. "An Education 21 Program: Orienting Environmental Education Towards Sustainable Development and Capacity Building for Rio." *Environmentalist* 15 (Autumn 1995): 202-10. EJ528318.

Presents a strategy and accompanying methodology for establishing environmental education as a major force for implementing Agenda 21. Proposes the establishment of an Education 21 program and the designation of the education community as a new Rio major group.

Richard, W. "The Globe: Neighbourhood Agenda 21, Going Local in Reading." *Streetwise* 5, no. 3 (1994): 5-9. EJ498188.

Reports on the philosophy underlying a project to promote local community involvement in neighborhood plans as a basis for city-wide Local Agenda 21 and the first stages of Go Local On a Better Environment (GLOBE).

Resources from ENC

Rescue Mission Planet Earth: A Children's Edition of Agenda 21. New York: Kingfisher, 1994. ENC-002855.

This book, written and illustrated by children, is designed to help young people understand Agenda 21, a document written as a result of the 1992 Earth Summit that outlines the course of action needed to better manage the world's growth and resources.

National Forum on Partnerships Supporting Education About the Environment. *Education for Sustainability: An Agenda for Action.* Washington, D.C.: U.S. Government Printing Office, 1995. ENC-011710.

This government document, developed by the President's Council on Sustainable Development, describes an action plan to integrate education for sustainability into broader curricula.

Tzelovanikov, exec. prod. *Ecological Design: Inventing the Future.* Videotape. New York: Ecological Designs Project, 1994. ENC-008813.

This video examines the emergence of ecological design in the 20th century. The film features the ideas and prototypes of designers who have pioneered the development of sustainable architecture, urban planning, and energy systems, as well as transport and in-dustrial engineering.

International Center for Conservation Education. *Caring for the World: A Strategy for Sustainable Living.* Slideshow. Cheltenham, Glos., U.K.: World Conservation Union, 1994. ENC-001117.

This 15-minute slideshow and script for grades four to 12 presents the background, principles, and key actions of caring for the Earth. This resource considers the ethical foundation for sustainable living and ways of translating principles into practical action.

Snyder, S.A., and Paden, Mary, eds. *Sustainable Development.* Wash-ington, D.C.: World Resources Institute, 1994. ENC-008331.

This teacher's guide, developed for grades nine to 12, is part of a series of units that integrate the global environment with science, mathematics, geography, history, and civics. Each unit introduces an issue and features practical lesson plans, teaching objectives, student handouts, discussion activities, suggestions for further readings, and audiovisual resources.

What Makes Education Environmental?

For many educators, environmental education is a subset of science. After all, it is the natural sciences that study the Earth, which is our environment. Yet most environmental educators rail against the supposition that environmental education is environmental science. Environmental education, they argue, is cross-disciplinary or transdisciplinary; it must include all studies — history, philosophy, economics, the sciences, mathematics, the arts, citizenship, and social studies — for after all, a decision on an environmental issue requires all facets of society to be considered.

If teaching environmental education is not teaching a particular set of theories supported by current facts within a framework we call a discipline, then what is teaching environmental education about? How would you know it if you saw it?

That is a surprisingly good question, with a not so readily apparent answer. Some practitioners in environmental education have suggested that "good environmental education is good education," referring to the commonly held beliefs in the field that environmental education is transdisciplinary-based teaching and experiential learning, learner engagement with the natural world, and critical thinking. But these types of learning are not ideals in just environmental education.

What *does* make education environmental if environmental education is neither science education nor everything else? The following articles provide some insight into the types of discussions and "best thinking" in the field regarding environmental education and what makes it unique.

Making the Experience of Environmental Education Experiential

by Joe E. Heimlich and Sabiha S. Daudi

A strength of environmental education is that it often engages the learner through exploration, examination, gaming, discovery, and hands-on activities. The nature of environmental education is such that these activities both draw the learner into the learning exchange and also become the basis for the learning that is to follow. By themselves, however, experiences do not necessarily teach learners. It is the role of the educator to help the learner frame knowledge from the impression, the game, the experiment, the discussion, the exploration, the hike, the animal, or whatever the experience. Let's explore what makes an experience in environmental education experiential learning.

What Is Experiential Learning?

When a visitor reads a sign in a nature reserve, does learning occur? What changes an observation into an understanding? At what point does a "fact" become knowledge? These are difficult questions and have no single or simple answer. Yet teaching and learning are predicated on the belief that there is the possibility of creating knowledge from information and sensory experiences.

Some theorists use the phrase "meaning making" to explain what occurs. Although an educator cannot make someone learn, the educator can structure an experience so that the potential for learning is enhanced. In experiential education, the process usually is described in a four- or five-step sequence:

1. Create the opportunity.
2. Involve the learner in an experience.
3. Process (discuss) the experience.
4. Generalize the experience to other situations.
5. Apply the knowledge.

These stages often are expanded or contracted and sometimes follow in a different progression. Yet writers on experiential education all tend to stress the similarities more than the differences. The dominant belief is that the experience does not teach, but prepares the learner for understanding the outcomes of the experience. The "learning" occurs in the processing and application that follow the experience.

Making Learning Experiential

Learning through an experiential process means taking a learner through each step of the cycle. Several tenets emerge in the literature that highlight some of the beliefs underlying experiential environmental education.

- An experience is unique for each learner.
- Each learner will react to an experience in a unique manner.
- Any experience can be used as a basis for learning, regardless of outcome.
- Individual learning preferences can be met in different stages of the cycle.
- The learner must create a framework for understanding the experience in a way that is relevant to prior experience and knowledge.

In environmental education, the experience is often given far more weight (and time) than the processing of the experience.

Critics of environmental education sometimes point out the lack of action or individual decision making that directly correlate to the application stage in the experiential cycle.

Because environmental education relies so heavily on experiences, the effort to create more opportunity for both processing and application is a logical extension to enhance learning. Such an effort can help make the experiences experiential.

To support environmental educators in their efforts to provide meaningful and experiential learning to the learners, we have identified some resources. These innovative approaches are necessary to link adventure and environmental education and suggest ways to enhance the processing of information gained through active encounters.

Resources from ERIC

Hanna, G. *Bridging the Gap: Linking Adventure and Environmental Education*. Alberta, Canada: EDRS, 1991. ED342589.

   This paper discusses ways to link components of the adventure education and environmental education system in Alberta, Canada. The paper list examples of teaching units emphasizing a theme, a metaphor, or a cognitive activity. The examples categorize activities into the foundation, exploration, and empowerment levels.

Wilson, Ruth A. "Nature and Young Children: A Natural Connection." *Young Children* 50 (September 1995): 4-11. EJ510603.

   This article discusses the value of nature education in early childhood education. The domains of adaptive, aesthetic, cognitive, communication, sensorimotor, and socioemotional development are explored. Guidelines, suggestions, and resources are offered for infusing nature education into early childhood education.

Silcox, Harry C. "Experiential Environmental Education in Russia: A Study in Community Service Learning." *Phi Delta Kappan* 74 (May 1993): 706-709. EJ463874.

   This article reports on the experiences of 26 U.S. high school students and environmentalists in Russia, where they participated in a community services and experimental learning environmental project.

Resources from ENC

Rowe, Mary B., ed. *What Research Says to the Science Teacher.* Vol. 1. Washington, D.C.: National Science Teachers Association, 1978. ENC-003578.

This first volume of the series examines the role of experiential learning in science, such as how to assess students' understanding of basic concepts from laboratory experiences. Teachers are invited to do field-based research to contribute to the data regarding outcomes of planned field experiences.

Sharing Success: Mathematics and Science Education. Tallahassee, Fla.: South Eastern Regional Vision for Education, Consortium for Mathematics and Science Education, 1993. ENC-001119.

This publication has information about professional development and reference material from the exemplary educational programs of SERVE. Among the many programs of excellence discussed are TACO (Take a Class Outdoors) and Using the Outdoors to Teach Experiential Science.

Braus, J.; Layton, P.; and Van Cleef, C. *Exploring Environmental Issues: Focus on Forests.* Project Learning Tree. Washington, D.C.: American Forest Foundation, 1995.

This Project Learning Tree (PLT) module uses forest issues as a focus for students to investigate and define an environmental issue, identify key players and points of view, generate alternative solutions, recognize and weigh trade offs, listen to their peers, and make personal and group decisions.

Environmental Education for Empowerment

by Francis E. Beasley, Joe. E. Heimlich, and Sabiha S. Daudi

Three decades after *Silent Spring* and two decades after Earth Day 1970, environmentalism often is still equated with wilderness and wildlife preservation, resource conservation, and the protection of endangered species. However, this concept of "environmentalism" is meaningless to many residents of inner-city communities threatened by other forms of environmental degradation, such as pollution, congestion, toxic wastes, and similar health-threatening influences. Lower-income, working-class, rural, and urban minority communities are disproportionately placed at risk by the more narrow concept of environmentalism.

The difficulty of attaining effective public policy objectives is a constant challenge to policy makers. Ensuring that the outcome of the policy objectives is also fair and equitable presents an even greater challenge. As we move into the 21st century, population shifts and demographic trends all point to a more diverse nation. The 1990s offered some challenging opportunities for the environmental movement to diversify and embrace an inclusive, rather than an exclusive, agenda that provides equal protection from environmental hazards to all persons.

Much of the debate between proponents and opponents of the concept of environmental discrimination is not centered on whether this discrimination occurs, but on whether the discrimination is driven by race or income. Limited income restricts the ability of poor people to move out of polluted communities and into environmentally safer surroundings. Land values are cheaper in these communities; thus industries seeking the most economical site often are attracted to such areas. In addition, environmental inequities result from other factors, including poor land use planning and lax enforcement of law. Minorities are disproportionately economically disadvantaged, and they are disproportionately affected by this type of development. Racial discrimination in education, employment, and housing are additional impediments faced by minorities in this country. In many cases, these added impediments contribute to the low pay and poor living conditions of the minorities.

Further impeding equity is that minorities usually are underrepresented in the decision-making process, thereby limiting their access to policy makers and to advocates for their interests. All these factors taken together cannot be overlooked. In addition, race has an additional effect on the siting of toxic release facilities — an effect independent of income.

In our capitalist society, history has played an important role in environmental inequities. The wealthy always have been better able to live "upstream or uphill," while those who work for them live below, where the water is less clean. The industrial revolution led to the wealthy being able to live away from the factories they owned or ran, while the poor needed to live nearby so they could walk to work. Since most people cannot escape environmental pollution, equitable action requires that the distribution of effects be shaped to be less harmful. Public policy ought to enable all in the city to enjoy the same high-quality environment wherever they are.

Environmental balance is an issue of equity. Environmental policy professionals, decision makers, and policy makers must address the outcomes or consequences of policies designed to ensure environmental protection for all.

By the same token, it is also the responsibility of the educator to involve learners in exploring important environmental issues of the community and in taking part in positive decision-making processes. Such education could include ensuring learner's understanding of the science and politics of saving our natural and cultural environments, of environmental justice, and of environmental decision making. Students empowered to solve problems in their own neighborhoods can help mobilize their communities to negotiate environmental issues and help them communicate their opinions to policy makers and community leaders.

To support the teacher in developing decision-making skills in their students, especially on the issues of environmental equity and empowerment, we have identified some educational resources that currently are available on the databases of Educational Resources Information Center (ERIC) and Eisenhower National Clearinghouse (ENC) collections.

Resources from ERIC

Victor, P. "About Ecology, Deep Ecology, and the Meaning of Life: A Talk to Teens." *Trumpeter* 11 (Spring 1994): 80-82. EJ491959.

    This talk on ecology and deep ecology addresses the interconnectedness of all things in the environment, the Gaia Hypothesis, as well as social inequity, lack of social and environmental harmony, and the adoption of globally sustainable lifestyles.

Bullard, R.D., and Wright, B.H. "Mobilizing the Black Community for Environmental Justice." *Journal of Intergroup Relations* 17 (Spring 1990): 33-43. EJ414370.

    Although black communities bear a disproportionate environmental burden because of institutional and locational factors, few blacks have joined the national environmental movement. Social justice and environmental equity are compatible goals, and the black community could take advantage of indigenous resources to develop strategies.

Bullard, R.D. "Overcoming Racism in Environmental Decision Making." *Environment* 36 (May 1994): 10-20, 39-44. EJ487140.

    Presents five principles of environmental justice to promote procedural, geographic, and social equity: guaranteeing the right to environmental protection, preventing harm before it occurs, shifting the burden of proof to the polluters, obviating proof of intent to discriminate, and redressing existing inequities.

McDonald, B. "Counteracting Power Relationships when Planning Environmental Education." *New Directions for Adult and Continuing Education* (Spring 1996): 15-26. EJ525527.

A case study of a planning effort in a low-income community describes how power was equalized between community activists and government planners. It shows that traditionally silenced voices need support to challenge authority, planners need ethical vision, and planning can be a learning process.

Resources from ENC

Dockterman, D. "The Environment: The Science and Politics of Saving Our Planet." *Decisions, Decisions Series.* Apple Software. Watertown, Mass.: Tom Snyder Productions, 1991. ENC-000497.

Decisions, Decisions is a series of role-playing software packages designed specifically to generate informed discussion and decision making in the classroom using only one computer. This package focuses on the environment and introduces science and social studies issues in the context of a local pollution crisis. The main topics include, among others, the role of government in achieving environmental quality.

"Balance and Decisions." *Science for Life and Living Series*, Level 6. Dubuque, Iowa: BSCS Innovative Science Education, Kendall Hunt. ENC-000448.

The Biological Science Curriculum Study (BSCS) Science for Life and Living Program uses a five-stage instructional model: engage, explore, explain, elaborate, and evaluate. The major concept is balance and the major skill is decision making. The students discover appropriate decision-making methods and options, interactions that occur within ecosystems, an awareness of the constraints and tradeoffs involved in solving problems, and different communication styles.

Inclusive Environmental Education

by Karen Ricker and Sabiha S. Daudi

How do K-12 educators in the classroom and in the field address the issue of making environmental education activities accessible to all students, even those with physical, developmental, and behavioral disabilities? Nationally, many school districts are moving toward the concept of "inclusion" for all students. This means that students with and without disabilities will be learning together in the same setting. The goal of inclusion is to have all students learn together, work together, and play and grow together. Inclusion also means: less exclusion; more students with disabilities in regular environments, more of the time, in more meaningful activities; having additional options for meeting the individual needs of students; and mainstreaming students, as well as offering support and resources. Inclusion and acceptance of challenged citizens into society is the goal of the Americans with Disabilities Act (ADA). Promoting and designing inclusive settings for environmental education activities, indoors as well as outdoors, will benefit all students and society.

The goal of education is to foster the development of lifelong learners, capable decision makers, and problem solvers, who will make valuable contributions to our society and to our world. With a finite and rapidly shrinking amount of natural resources on our Earth, coupled with an expanding rate of population growth, it is imperative that we foster and facilitate learning in and about our environment with each and every student, regardless of ability, if we are to reach this goal.

Environmental education activities can promote a student-centered approach to learning. As a facilitator of learning, teachers can help students actively participate in hands-on experiences by selecting activities that relate to the curriculum or the student's individualized education program (IEP) goals. Experiential, hands-on activities also offer a different learning modality for students than the classroom. For the special education student who has been labeled "learning disabled," it may mean a whole new way of learning. The classroom may not be the "least restrictive environment" for students in terms of learning styles — that may be the outside environment.

Does this mean that educators need to spend more time developing additional environmental education activities for students with disabilities? No, not at all. There is no reason to "re-invent the wheel" by developing new programs when many outstanding environmental education programs already exist. The only requirement may be the need to modify the activities in order to meet students' needs so that they may fully participate in the activities.

Some Pointers for Adapting Activities

When adapting activities for inclusive settings, we need to remember that each individual is unique. Here are some pointers to be kept in mind when adapting or developing activities for inclusive learning.

- Prepare participants without disabilities by giving them general information about disability issues; specific information related to the needs of participants with disabilities, the opportunity to role play by having students "try out" the disability, and such guidelines as discussing the role of all participants, those with and without disabilities, to increase participation of all group members.

- Include participants with and without disabilities in decision-making processes regarding adaptations, rules, and other changes.
- Remember that the participant with a disability knows his or her abilities better than anyone else does; nondisabled peers will feel included by helping to make decisions about modifications; and activities may be more successful because all participants were involved in initial adaptation decisions.

Resources from ERIC

Newbury, M. "An Environmental Approach to Pupils with Special Needs." *Environmental Education* 45 (Spring 1994): 30-31. EJ480089.

Describes the interdisciplinary approach of one British school for challenged students to make students aware of environmental issues and to foster active participation in the care of the school and the extended environment.

Schlein, S.J., et al. "Integration and Environmental/Outdoor Education: The Impact of Integrating Students with Severe Development Disabilities on the Academic Performance of Peers Without Disabilities." *Therapeutic Recreation Journal* 28, no. 1 (1994): 25-34. EJ485718.

Reports a study that assessed the amount of environmental information acquired by nondisabled children while participating in a one-day integrated outdoor education experience with severely developmentally delayed children.

Filmer, R. "Environmental Education and the Disabled." *Environmental Education Bulletin* (November 1990): 26-32. EJ449293.

Describes specific plans for making environmental education accessible to disabled people by adapting trail systems and facilities for safety, interpretative programming, and access.

Resources from ENC

Torres, I., and Corn, L.A. *When You Have a Visually Handicapped Child in Your Classroom: Suggestions for Teachers.* New York: American Foundation for the Blind, 1990. ENC-000352.

This booklet is intended as a source of information and guidance to regular classroom teachers in elementary and secondary schools who have visually impaired children in their classroom. The teacher is introduced to the special education needs of the blind and low-vision students. Information on expectations, orientation, mobility, and social skills is included and can be useful when adapting activities for environmental education.

Environments. Berkeley, Calif.: SAVI/SELPH, Center for Multi Sensory Learning, Lawrence Hall of Fame, 1983.

The SAVI/SELPH program is an interdisciplinary, multi-sensory, science-enrichment program designed for physically, emotionally, and mentally challenged students in grades four to seven. This module explores environments and has activities on environmental planting, in which students grow seeds and bulbs under controlled conditions and observe the results.

Environmental Energy. Berkeley, Calif.: SAVI/SELPH, Center for Multi Sensory Learning, Lawrence Hall of Fame, 1983.

This module explores environmental energy and has activities on solar energy and wind power, which introduce students to the concepts of solar energy and energy transfer.

Using Community Resources in Environmental Education

by Emily Ciffone, Jolin Morelock, Karen Turner, Dan Sivek, and Sabiha Daudi

Community resources are the people, places, or objects located off the primary site (usually the school) that may be used to achieve educational objectives. Environmental education lends itself quite naturally to community-oriented teaching. The use of resources within a community can greatly enhance and expand the school curriculum. Community resources can help teachers teach more effectively by providing motivation to students, helping students achieve learning objectives, and exposing students to positive role models and "real life" situations.

Community resources can provide the motivation some students need to see the connection between the classroom and the real world. Involving students in the community gives them exposure to a stimulating learning environment and to different people and perspectives and provides students with a greater sense of purpose.

Often community-based activities can help students fulfill desired learning outcomes in a manner that is more engaging than traditional textbook assignments. For instance, a middle school science curriculum may include these learning objectives: "Develop an understanding of the recycling process," "Distinguish between number-one and number-two plastics," and "Read and construct a bar graph." In a carefully planned field trip to a recycling plant, students can gain relevant firsthand experience with those concepts and processes.

The use of community resources can further the goals of environmental education by preparing students for the real world and helping students to become "world class" citizens. With such a mindset, the community becomes an extension of the school. Since all environmental problems occur in someone's community, it seems logical for students to try to solve local issues. The teacher can facilitate this process by choosing manageable initial projects with a high chance of success. In turn, students will begin to see themselves as problem solvers and will likely continue their involvement. As the students work on solutions, they will connect with caring local citizens who are affected by environmental problems. In so doing, they will develop a sense of stewardship and place, which will help build the bridge to global responsibility.

Implementing Community Resources

Community resources in the form of people, places, or things can be found in all sectors of the community and can provide teachers with teaching materials (borrowed or donated), project ideas, guest speakers, field trips, or community service projects. Some organizations hold workshops, which inform teachers about various service, financial, equipment, or curricular resources that they offer. Every community, no matter how large or small, holds cultural, natural, human, and technological resources that can be used by the students and teachers who live there.

Three common ways for educators to use community resources in environmental education include: guest speakers, field trips, and community projects. Guest speakers can be anyone from environmental professionals to students' family members to other community members. Possible field trip destinations are zoos, manufacturing plants, farms, or other community businesses where students can interact with the people who work there.

These businesses also may serve as sources of internships or mentoring experiences for students. Community service projects can take many forms, from picking up litter to long-term adoption of a natural area.

Learning through community resources can involve short classroom sessions, one-day field trips, a week-long unit, or a year-long project. Lessons can be interdisciplinary or focus in-depth on a particular topic. Once educators begin using a variety of community resources, it may be best to begin an inventory of those resources. An inventory can take the form of a notebook, a box of index cards, or a computer database. Each entry may have the name of the resource, the appropriate subject areas, what the resource can be used for, contact name, address, and a list of restrictions or limitations pertaining to the resource. This catalogue of community resources allows several educators to share their ideas and experiences with others in their school.

Barriers to Implementation

There are a variety of reasons educators may be reluctant to use community resources to supplement their environmental education instruction. Listed below are examples of potential barriers along with some suggestions of possible solutions:

1. Time commitment involved. Encourage student participation in all aspects of the project. Use community resources in place of text material. Select an issue or project that fulfills several learning objectives. Join with other classes to increase the amount of class time devoted to a project.
2. Additional costs. Consider all costs before proceeding with the project. Find organizations that provide free materials. Consider the use of volunteers (high schools and universities are good sources). Design activities that can be completed at or near school to decrease transportation costs. Develop and implement fundraising strategies with the students.
3. Discipline/safety issues. Be prepared. Consider the factors that may cause problems. Make rules and expectations clear

before the event. Enlist students to help monitor behavior. Ask for parental help in chaperoning, if necessary.

4. Inexperience in using community resources. List the learning objectives that must be covered and list possible community resources. Match potential resources to the objectives. Enlist the help of other teachers. Many teachers do use community resources in their classrooms and can provide you with invaluable advice and contacts.

5. Lack of support. Become familiar with school policies before confronting administrators. Consider enlisting their help before you plan an event in order to increase their participation in the process.

Resources from ENC

Science Framework for California Public Schools: Kindergarten Through Grade Twelve. Sacramento: California Department of Education, 1990. ENC-001424.

This is a curriculum framework document from California that is designed to increase the connections among science disciplines and technology and society. The document deals with the use of educational technology in classrooms, the physical resources of the school and the community resources of the district, and guidelines for staff development.

The Outdoor Classroom: Experiencing Nature in the Elementary Curriculum. Indianapolis: Indiana Department of Education, 1989. ENC-00100.

This document provides 10 lessons using outdoor education, hands-on learning field trips, and an interdisciplinary approach to explore a variety of subjects for grades K-6. Through direct experience, students learn about wildlife, weather, habitats, watersheds, erosion, geology, food chains, forestry resources, harvesting crops, and community resources.

Resources from ERIC

Nature Education in the Urban Environment: Proceedings of the Forum. New York: Bank Street College of Education, Central Park Conservancy, and Roger Tory Peterson Institute of Natural History, 1991. ED347029.

This conference report answers the question of how to encourage the more positive use of parks for outdoor nature education. Recommendations include improving teacher training with respect to nature education, developing school-community partnerships, and acknowledging the importance of camping, outdoor experiences, and individualized curricula in nature education.

Mathews, Bruce. "Building Connections: Remarks at the Tenth Annual Conference of the Michigan Association for Environmental and Outdoor Education." *Taproot* 1 (September 1997): 6-8. EJ554507.

The author emphasizes that environmental educators must rebuild a sense of place by building personal stories in particular places over time. These stories impel us to care about the places where the stories occur, thus creating responsible environmental stewards.

Environmental Education and Nature Centers

by Tara Tucker, Sarah Kiser, Dan Sivek, and Sabiha S. Daudi

According to the American Association of Museums (as found in Natural Science Centers Directory), a nature center is

> an organized and permanent nonprofit institution which is essentially educational, scientific, and cultural in purpose with professional staff, open to the public on some regular schedule. The Nature Center manages and interprets its lands, native plants and animals and facilities to promote an understanding of nature and natural processes. It conducts frequent environmental education programs and activities for the public.

Due to the diversity among nature centers, it is impossible to offer a description of a "typical" nature center.

The Natural Science Centers Directory and a study of 1,225 nature and environmental education centers by Deborah Simmons (1991) help define characteristics of the majority of nature centers. Approximately 80% of centers provide programs for day users only. The remainder include some residential programming in their offerings. The majority of nature centers are run by government agencies, though approximately 40% are run by private nonprofit organizations. Nature centers serve urban, suburban, and rural audiences. The Simmons study found that

95.3% of nature and environmental education centers served elementary school groups.

How Nature Centers Complement Environmental Education

The ultimate aim of environmental education is responsible environmental behavior. The goals essential in achieving this aim include sensory awareness, knowledge, values, citizen action skills, and citizen action participation. From preschool through 12th grade, environmental education engages children in lessons that enforce these goals. Nature centers can supplement teachers' lessons by offering hands-on experiences that children can only read about in the classroom.

In particular, nature centers are useful in establishing a connection to the environment. With an increasing human population, society is becoming more urbanized. The U.S. population is growing at an annual rate of 0.9%, and 75% of the population is urban (Population Reference Bureau 1998). Thus, fewer children spend time catching frogs in ponds, inspecting critters in the soil, and listening to the wind blow through a dense stand of aspen trees. Without a connection to the outdoors, the goals of environmental education may not be achieved.

Most nature centers focus on developing awareness and knowledge of the natural environment. The Simmons study (1991) revealed that 75% of the nature and environmental education centers surveyed promoted nature study and environmentally sound behavior. Fewer centers promoted citizen action skills (40%) and attitudes/values (66%). This difference was attributed to the reluctance of nature centers to get involved in controversial environmental issues. Although not all nature centers strive to meet every environmental education goal, progress has been made. In her 1998 epilogue to the nature center study, Simmons states, "we need to start thinking about how comprehensive programs can be built using all of the resources available to the school. Nature centers, zoos, and museums can play an important, if not essential role, in this process" (p. 319).

The focus on awareness and knowledge of the natural environment also may be useful in dispelling fears students may have about the outdoors. A 1994 study (Bixler et al. 1994) revealed that fears of snakes, insects, getting lost, poisonous plants, and getting dirty are experienced by a high percentage of visitors to nature centers. These fears can impede students' learning in a natural environment. By offering programs that help students explore the outdoors in a structured environment, nature centers may help alleviate these fears.

To achieve maximum benefits from a nature center visit, teachers should properly prepare students prior to reaching the center. Watson found that students will not receive new information effectively unless they have been acclimated to the subject matter and center before instruction takes place. Pre-trip activities that relate to the field trip may be conducted in the classroom. This helps direct students towards the focus of their visit, and learning at the site will be enhanced. Some examples may include writing in journals or drawing what they expect to see at the nature center. Post-trip lessons are essential in reinforcing the concepts presented during the students' visit. The most successful pre- and post-visit activities are those developed by nature center staff in collaboration with teachers (Paris 1994). These activities should complement the nature center lesson to promote maximum understanding of the concepts taught.

References

Bixler, R.D.; Carlisle, C.L.; Hammitt, W.E.; and Floyd, M.F. "Observed Fears and Discomforts Among Urban Students on Field Trips to Wildland Areas." *Journal of Environmental Education* 26, no. 1 (1994): 24-33.

Population Reference Bureau. "U.S. in the World: Connecting People and Communities." 1998. http://www.prb.org/prb/news/usworld/overview.pdf

Simmons, D.A. "Are We Meeting the Goal of Responsible Environmental Behavior? An Examination of Nature and Environmental Education Center Goals." *Journal of Environmental Education* 22, no. 3 (1991): 16-21.

Simmons, D.A. "Epilogue or Some Future Thoughts on: 'Are We Meeting the Goal of Responsible Environmental Behavior? An Examination of Nature and Environmental Education Center Goals'." In *Essential Readings in Environmental Education*. Champaign, Ill.: Stipes, 1998.

Resources from ERIC

Touvell, D., et al. *Natural Science Centers: Directory*. Roswell, Ga.: Natural Science for Youth Foundation, 1990. ED319619.

This directory serves as a comprehensive list of nature centers in the United States and Canada. It includes zoos, centers, parks, and museums. Overall statistics include audience use, management, and budget of the 1,261 centers.

Paris, Rhana Smout. "Bring Them Prepared: Developing Pre-Trip and Post-Trip Lessons for Visiting Students and Their Teachers." *Legacy* 5, no. 3 (1994): 30-32. EJ491866.

This article discusses the need of pre-trip and post-trip activities in creating a more meaningful experience at nature centers. Includes an outline for activity development and emphasizes teacher-nature center staff collaboration in creating activities.

Butts, David S. "Blazing a Nature Trail — Behind the School." *Principal* 64, no. 3 (1985): 38-41. EJ323572.

An example of how nature centers can work in tandem with schools to provide programming on the school site. A community nature center provides resources, instruction, and teacher inservice training to help third-graders build a nature trail for the community on their school grounds.

Internet site related to nature centers. "Science Adventures: The Guide to Informal Science Centers." www.scienceadventures.org

A guide to finding informal science education centers throughout the United States. Includes museums, zoos, nature centers, parks, planetariums, and observatories.

Resources from ENC

Ballbach, J., ed. *Ohio Sampler: Outdoor and Environmental Education*. Newark: Environmental Education Council of Ohio (EECO), 1994. ENC-005882.

This book contains environmental and outdoor education activities for K-12 students. The activities represent personal favorites of environmental and outdoor educators who have used them with their own students.

Warren, K.J., and Andrews, E. *A Guide to Unique Program Strategies.* Columbus, Ohio: ERIC/CSMEE, 1995. ENC-010593.

This resource guide, developed for educators as part of the Educating Young People About Water series, helps water education program coordinators design a program strategy appropriate for their local situation.

Integrating Environmental Education and Formal Education

Environmental education is not considered a discipline, so it is not historically a part of study within the formal education system. However, many educators find that environmental education has a valuable role within the formal education system. Likewise, many of the institutions or organizations that traditionally offer environmental education or interpretive programs have discovered that partnerships with schools provide a captured audience and a way to reach many more youth with messages about their resources or environmental perspectives.

Many of the environmental education organizations or providers, whether nongovernmental agencies or nature centers, zoos, parks, aquaria, and the like, are increasingly dependent on grants to support and expand their educational programs. One perception is that many grants require large numbers of students, usually thought of as children in classrooms, to justify funding. Another competing perception is that partnering with schools benefits schools with programming and the environmental education organization with visibility.

There long has been both tension and celebration in the relationship of environmental education to formal education. Examining the journals of the last few decades reveals the cyclical nature of the discussions regarding the relationship of environmental education to formal education: separation, integration, thematic based, transdisciplinary, and other such ideas cycle around, seem to disappear or at least become internalized, and then emerge again several years later in a slightly different configuration. This section presents articles that discuss some resources and approaches to integrating environmental education into existing classroom curricula.

Environmental Education: A Tool for Making Education Reform Work

by Michele Archie, Joe E. Heimlich, and Sabiha S. Daudi

Many of the goals of environmental education and education reform are strikingly similar. These include helping students to be knowledgeable and skilled thinkers who are able to put their knowledge, skills, and creativity to work solving problems, who are practiced at working collaboratively and independently, and who are prepared to take their role as responsible citizens. Yet there has been little collaboration between the educators and organizations that promote the two endeavors.

To foster dialogue and collaboration among the environmental education and education reform communities, the EdGateway website at www.edgateway.net suggests several areas of overlap between the interests of the two. EdGateway is an Internet-based information resource for educators and education organizations and is sponsored by the WestEd Eisenhower Regional Consortium for Mathematics and Science Education. These areas of convergence form the basis of the following discussion of environmental education as a tool for making education reform work.

Curriculum

Many reform efforts focus on the curriculum — the knowledge and skills we want students to learn. In large measure, the current education reform movement hinges on setting standards for learner achievement. Voluntary national standards have been put in place for disciplines from science to English language arts to fine arts to mathematics, and many states have adopted their own standards for student achievement.

Environmental education can be a tool for meeting these standards. *Excellence in Environmental Education — Guidelines for Learning (K-12)*, published in 1998 by the North American Association for Environmental Education (NAAEE), makes these links explicit. Developed through a national process of review and comment involving more than 2,500 individuals and organizations, *Excellence in Environmental Education* covers the core concepts and skills that environmentally literate citizens need. Guidelines and performance measures are suggested for the fourth, eighth, and 12th grades. Each guideline is linked to related discipline-based standards.

Excellence in Environmental Education is based on a vision of a curriculum that is in keeping with education reform goals: emphasizing higher-order thinking skills over memorization and repetition, linking to the world outside the classroom, and using the environment as an integrating theme that links disciplinary studies together. There are several model efforts under way to use the environment as an integrating concept in schools around the nation. For example, there are descriptions of 10 school prototypes integrating curriculum around environmental education at http://cisl.ospi.wednet.edu/CISL/ENVED/MDLINKS.html ("Creating Model Links: Environmental Education and Education Reform in Washington").

Instruction

Education reform also targets instruction — the ways in which we expect teachers to teach and students to learn. The emerging vision of instruction crosses disciplinary lines, is hands-on, guided by student interests and responsibility, and often collaborative. Based on a similar instructional vision and supported by hundreds of high-quality instructional materials, environmental education offers tools for implementing education reform.

Environmental Education Materials: Guidelines for Excellence, a 1996 publication of NAAEE, recommends guidelines for selecting, evaluating, and producing top-quality environmental education lesson plans, curricula, and other instructional materials. Using these guidelines, educators and content specialists reviewed hundreds of educational materials; and NAAEE published their reviews in four resource reviews designed to help educators identify resources to meet their needs.

The instructional vision that environmental education and education reform efforts share requires a new level of professional development from teachers. NAAEE's *Guidelines for the Initial Preparation of Environmental Educators* (2000), offers a set of recommendations about the basic knowledge and abilities educators need in order to provide high-quality environmental education.

The School Site

Education reform must be successful at each individual school if it is to work at all. There are many reform efforts that focus on changing how the school site itself is used and how it relates to the larger community. Environmental education offers the opportunity to use the environment as a learning context — starting from the school grounds and extending out to the community and beyond. This approach could be as simple as students conducting an environmental audit of the school and school grounds. Or it could be more comprehensive, such as an environmental "magnet school" or a school that uses the environment as an integrating concept. Environmental education also promotes parent and community involvement.

Assessment

If we change our ideas about what students should learn and the kind of instruction that will best help them do that, then we'll also need to change how we measure student success. Education reform efforts promote alternative assessments, such as perform-ance assessments, that tend to measure whether students have mastered and can apply knowledge and skills — rather than sim-ply perform rote techniques or repeat memorized facts.

Environmental education lends itself to using student projects, portfolios, and other tools that assess a wide range of skills. *The Guidelines for Learning* include sample performance indicators for each guideline at the fourth-, eighth-, and 12th-grade levels. These illustrate some ways in which learner achievement might be demonstrated. And assessment is one of six thematic areas in NAAEE's forthcoming guidelines for the preparation of environ-mental educators.

Resources from ERIC

O'Neal, E.C., and Kirk, P.A.C. "Effects of Educational Reforms on Pre-Post Reform NTE Scores of Physical Education Majors." Paper presented at the Annual Meeting of the Mid-South Educational Research Association, Knoxville, Tenn., 11-13 November 1992. ED353325.

   After a review of the recommendations of national reform reports in the 1980s concerning teacher education programs, legislation, and regulatory systems established in Mississippi to facilitate education reforms, this paper describes a study of the American College Test, which predicts academic success in college, and the National Teach-er Examination (NTE), which measures the knowledge base for teaching and predicts academic competency.

Cole, A.L.; Elijah, R.; and Knowles, G.J., eds. *The Heart of the Matter: Teacher Educators and Teacher Education Reform*. San Francisco: Caddo Gap, 1998. ED425153.

   This collection of papers examines the role of teacher educators in teacher education reform. Divided in four parts, Part 1 focuses on the reform context, Part 2 on self-study as teacher education reform, Part 3 on teacher educators and the reform of teacher education, and Part 4 on deans of education and college reforms.

Resources from ENC

Futrell, M.H.; Lynch, S.; and Hinter-Boykin, H. "Reaching All Stu-dents of Diverse Needs and Cultures." 1996. ENC-007671.

This paper cites methods and curriculum materials that can be used by teachers to implement science and mathematics education reforms while accommodating students of diverse needs and cultures. Full text of this document can be accessed at http://www.enc.org/reform/journals/102884/2884.htm.

Environmental Education In a Standards-Based Curriculum

by Michele Archie, Bora Simmons, Joe E. Heimlich, and Sabiha S. Daudi

In large measure, the current education reform movement hinges on setting standards for learner achievement. The 1991 federal education reform act, variously known as Goals 2000 or America 2000, mandated standards-setting efforts for the traditional disciplines. In the ensuing years, voluntary national standards have been put in place for disciplines from science to English language arts, and from fine arts to mathematics and social studies.

Across the country, states and school districts are beginning to hold teachers accountable for their students meeting these standards or other benchmarks set at the state and local level. One effect of this accountability is that curricula are being adjusted so they are more likely to lead students to perform well on standards-based evaluations. Another effect is that teachers are feeling more pressed than ever to find time for new content and teaching activities within a crowded curriculum.

Environmental education often is seen as one of the "add-ons" to a set curriculum or as just something else to find time for in a school day that already is filled to capacity. However, supported by hundreds of high-quality instruction materials, environmental education can and does "fit" in the curriculum. The trick is to make the link between the traditional disciplinary standards and environmental education.

Each of the voluntary national standards for traditional discipline areas does, to one degree or another, incorporate learning and instructional goals that are aligned with those of environmental education. For example, essential ecological knowledge is included within the geography standards. Similarly, understandings of measurement, patterns and relationships, and statistics and probability, all elements of the standards for mathematics, also are important to environmental literacy.

Environmental education programs can be used to teach a variety of concepts and skills within disciplines. And because environmental education is, at its heart, an integrative undertaking, it can help educators and learners cross disciplinary boundaries. Understanding environmental connections requires that students are able to link methods and ideas from natural and social sciences, arts, mathematics, and humanities. Learning about the environment and environmental issues is a continuing lesson in interconnectedness that draws on the core disciplines and provides a meaningful context.

Until recently, there's been no comprehensive means of showing educators how they can develop programs that meet the needs of both the core curriculum and environmental education. *Excellence in Environmental Education: Guidelines for Learning (K-12)*, published in 1998 by the North American Association for Environmental Education, makes these links explicit. Developed through a national process of review and comment involving more than 2,500 individuals and organizations, *Excellence in Environmental Education* provides a shared view of the core concepts and skills needed by environmentally literate citizens. Guidelines and performance measures are suggested for the fourth-, eighth-, and 12th-grade levels. Each guideline also is tied to related discipline-based standards.

Excellence in Environmental Education provides the scaffolding on which cohesive, sequential, comprehensive, environmental education programs can be created. Taking a holistic approach to environmental education can create synergy among its basic parts — or subject areas. As students analyze and evaluate the complexities of an environmental issue, they begin to understand intricacies and connections they could not have discovered if the information was presented fact by fact and subject by subject outside the context of the environment as a whole.

Only the development of a comprehensive environmental education program insures that it will not be marginalized or fragmented. To be effective, these programs must be constructed with a clear understanding of the knowledge and skills that lead to environmental literacy and with a vision of environmental education's place within the school curriculum. Other resources are available to help educators create environmental education programs that fit within and further the effectiveness of a standards-based curriculum.

Resources from ERIC

Simmons, Deborah, et al. *Environmental Education Materials: Guidelines for Excellence*. Washington, D.C.: North American Association for Environmental Education, 1996. ED403145.

These guidelines are a set of recommendations for developing and selecting environmental education materials, with the aim of helping developers of activity guides, lesson plans, and other instructional materials produce high-quality products and of providing educators with a tool to evaluate the wide array of available education materials.

Fortier, J.D.; Grady, S.M.; Susan, M.; Lee, S.; and Marinac, P.A. *Wisconsin's Model Academic Standards for Environmental Education*. Bulletin No. 9001. Madison: Wisconsin State Department of Public Instruction, 1998. ED426855.

This guide to Wisconsin's academic standards for environmental education describes the process and development of state environmental standards. Designed for administrators, school board members, and teachers, the guide explains the purpose and goals of creating standards and contains a brief history of environmental education in Wisconsin.

Division of Applied Technical, Adult and Community Education. *Agribusiness and Natural Resources Education: Vocational Education Program Courses Standards*. Tallahassee: Florida State Department of Education, 1997. ED409428.

This document contains vocational education program course standards (curriculum framework and student performance standards) for exploratory courses.

Green, D., et al. *The School Ground Classroom: A Curriculum to Teach K-6 Subjects Outdoors*. Portland: Environmental Education Association of Oregon, 1980. ED219286.

Includes lesson plans and activities to demonstrate that the outdoors is an interdisciplinary classroom that can be used on any school site and that can teach subject matter taught as part of the standard curriculum.

Resource from ENC

Conley, D.T. *Are You Ready to Restructure? A Guidebook for Educators, Parents, and Community Members*. Thousand Oaks, Calif.: Corwin, 1996. ENC-006065.

This book is designed for individuals wanting to make fundamental changes within their school. It is intended for administrators, teachers, school board members, parents, students, and other community members. It aims to provide these individuals with information that equips them to make more effective decisions for their schools.

Project WILD:
An Experience in
Experiential Education

by Sabiha S. Daudi and Joe E. Heimlich

Experiential education is based on the premise that the outcomes of teaching and learning cannot be predicted solely on the basis of providing facts and information about any scientific or nonscientific concept for the purpose of creating knowledge. As described in the article, *Making the Experience of Environmental Education Experiential*, the process usually is described in a multi-step sequence:

1. Create the opportunity.
2. Involve the learner in an experience.
3. Process (discuss) the experience.
4. Generalize the experience to other situations.
5. Apply the knowledge.

These stages often are expanded or contracted and sometimes follow in a different progression. Yet writers on experiential education all tend to stress the similarities more than the differences. The dominant belief is that the experience does not teach but prepares the learner for understanding the outcomes of the experience. The "learning" occurs in the processing and application that follow the experience.

Project WILD

This is also the mission of a unique experience in environmental education, Project WILD, which aims to provide wildlife-based environmental education that fosters responsible actions toward wildlife and related natural resources. The main goal of Project WILD is to assist learners of any age in developing awareness, knowledge, skills, and commitment that will result in informed decisions, responsible behavior, and constructive actions concerning wildlife and the environment. This directly ties in and is in accordance with the charter given to environmental educators at the International Environmental Education Workshop held in Belgrade in October 1975.

Project WILD is an interdisciplinary, supplementary, conservation and environmental education program emphasizing wildlife. Project WILD's primary audience is educators of students from kindergarten through high school.

Background

Project WILD is sponsored by the Council for Environmental Education (CEE), a nonprofit organization composed of representatives from state departments of education and natural resource agencies from all 50 states. CEE, formerly the Western Regional Environmental Education Council, has been the sponsor of the program since 1973.

In 1992, with support from the National Fish and Wildlife Foundation and Phillips Petroleum Company, the Project WILD Action Program was launched to inspire ideas and provide models for conducting effective environmental projects that dynamically engage students from start to finish. Later, in cooperation with the National Wildlife Federation, the WILD School Sites guide was published to help students and teachers learn about the importance of biodiversity and understand the basic steps of creating a wildlife habitat on school grounds. A complementary video, "Exploring School Nature Areas," was produced in cooperation with St. Olaf College's "School Nature Area Project" to inspire students and educators to take positive action on behalf of the environment.

In 1995 Project WILD completed an intensive long-range planning process that resulted in the establishment of goals and objectives through the year 2000. WILD in Service, an environment-focused service learning initiative, is central to three of these goals: integrating the concept of responsible action into all programs; increasing effectiveness with underserved audiences, including secondary schools; and seeking partnerships in the development and implementation of new programs. The program was piloted in Maryland and Texas, with support from the National Environmental Education and Training Foundation (NEETF) and Phillips Petroleum. Project WILD is also a partner in the Environmental Education and Training Partnership (EETAP).

Audience

Project WILD is available on a statewide basis in those states where the state wildlife agency, working in cooperation with the state's department of education, sponsors the program. Project WILD currently is available in all 50 states, the District of Columbia, Puerto Rico, Canada, the Czech Republic, Iceland, India, and Sweden. Project WILD is the most widely used environmental education program in North America. Since the fall of 1983, more than 600,000 teachers have attended Project WILD workshops, reaching more than 35 million students.

Modus Operandi

The Project WILD materials consist of two interdisciplinary activity guides for educators: the *Project WILD K-12 Activity Guide*, emphasizing terrestrial wildlife and habitat, and the *Project WILD Aquatic Education Activity Guide*, emphasizing aquatic wildlife and ecosystems. The Project WILD materials are available to all educators, formal and nonformal, through their participation in a Project WILD workshop. The conceptual framework for Project WILD was developed through a rigorous process to ensure its accuracy, balance, and educational validity. It was critiqued and reviewed by more than 500 professionals.

Project WILD uses effective experiential methods to teach problem solving and decision making. It teaches students "how to think, not what to think." Project WILD guides the student through a process that begins with awareness, moves toward understanding, challenges preconceived notions, and instills the confidence, skills, and motivation to take responsible action on behalf of the environment. Each Project WILD activity lists instructional objectives and specifies the intended learner outcomes. The activities are designed to be used singly, sequentially, or in clusters with youth in classrooms or outside the formal classroom setting.

Evaluation

Another major goal of the long-range plan is the summative and formative evaluation of all Project WILD program activities to show the results of past programs and to reflect a client-oriented approach to shaping future programs and revising existing programs. The dual objectives of this evaluation are to complete an integrated evaluation of the effectiveness of Project WILD and to develop a strategy for ongoing evaluation at both the national and state levels. The evaluation design developed by the Institute for Learning Innovation (ILI) consists of two major phases: an impact study and a market analysis. The evaluation will be conducted over three years.

Through its network of state sponsors, primarily fish and wildlife agencies, Project WILD often is adapted or expanded to ensure that teachers and students learn about wildlife and environmental issues unique to their geographic area. Internationally, Project WILD works to identify a major sponsoring organization in each country, which must license the right to use, adapt, or translate Project WILD materials. Since Project WILD materials are copyrighted, permission to reprint or adapt the materials must be granted through the Project WILD national office.

For more information about Project WILD, check out their web site at www.projectwild.org. Or contact the Project WILD National Office at (301) 527-8900 or by fax at (301) 527-8912. The address is 707 Conservation Lane, Gaithersburg, MD 20878.

Project WILD also has been evaluated and adopted by many leading environmental organizations. Some of these reports are available in the databases of Educational Resources Information Center (ERIC) and Eisenhower National Clearinghouse (ENC) collections.

Acknowledgement: The authors thank Donna Asbury and Gwyn Rozzelle of Project WILD and Josetta Hawthorne of CEE.

Project Learning Tree:
Hands-On Learning in
Environmental Education

by Joe E. Heimlich and Sabiha S. Daudi

Experiential education is the process of actively engaging students in an experience that will develop knowledge and understanding of "real life" issues with consequences. Students make discoveries and experiment with knowledge themselves, instead of hearing or reading about the experiences of others. Students also reflect on their experiences, thus developing new skills, new attitudes, and new theories or ways of thinking. One approach to fortify this concept is to develop curricular resources that strengthen learning processes by providing information, activities, and firsthand experiences based in learning situations and challenges relevant to the learner. Environmental issues have been successfully addressed in the classroom by educators to promote critical-thinking and problem-solving skills in learners through experiential education.

Project Learning Tree

Project Learning Tree (PLT) is an example of such a successful effort. It is designed to work in rural, suburban, and urban areas, whether there is a forest, a single tree, or a vacant lot. The PLT activities emphasize conceptual learning and skill building and use effective, student-centered instructional strategies, such as hands-on and cooperative learning.

Project Learning Tree (PLT) is an award-winning, interdisciplinary, pre-K-12, supplementary, environmental education program. PLT has a pre-K-8 guide and a series of secondary level modules. These modules include such topic-specific issues as Forest Issues, Forest Ecology, Municipal Solid Waste, and Risk. PLT uses readily available resources from the natural and built environments to help students gain awareness and knowledge of the world around them and their place within it, as well as their responsibility for it. Besides increasing students' understanding of their environment, PLT's mission is to stimulate students' critical and creative thinking and instill in students the confidence and commitment to take responsible action on behalf of the environment. PLT uses the forest as a "window on the world" to increase students' understanding of our complex environment and to develop the ability to make informed decisions on environmental issues. Project Learning Tree is sponsored nationally by the American Forest Foundation, a charitable education foundation.

Audience

PLT is one of the most widely used pre-K-12 environmental education programs in the United States and abroad. Since its inception in 1973, more than 500,000 educators have attended training workshops and have, in turn, reached more than 20 million students in all 50 states and the District of Columbia, several U.S. Territories, Canada, Sweden, Finland, Japan, Brazil, Chile, and Mexico. PLT trains more than 30,000 educators every year.

As a grassroots volunteer program that works in conjunction with teachers, schools, state agencies, business and civic organizations, museums, nature centers, and youth groups, the program is guided by state steering committees, led by a network of 100 state coordinators, and supported by more than 3,000 volunteer facilitators.

Modus Operandi

PLT can be used in both formal and nonformal education settings, including schools, community programs, museums, and nature centers. Project Learning Tree provides workshops and inservice programs for educators in grades pre-K-12, preservice students, foresters, park and nature center staff, youth group leaders, home-schoolers, and other community members interested in using the program.

PLT motivates educators to take part in the program by providing a two-tier training approach and through its innovative materials. The first tier is the facilitator training workshop, in which volunteers are trained to conduct educator workshops. Educator workshops serve as the second tier of the training component. The curricular materials are distributed at educator workshops and are not available commercially. These workshops give guidance on the use of the PLT materials, increase educators' understanding of how to teach environmental issues, and allow educators to share important teaching experiences. PLT training workshops improve educators' confidence and skills in teaching environmental education and give them the opportunity to network with other educators, resource managers, and other professionals who can assist with teaching about environmental issues.

Evaluation

All of PLT's curriculum materials have gone through both formative and summative evaluations. In 1993 and 1994, PLT took on the task of evaluating their curriculum. PLT hired the North American Association of Environmental Education's Commission on Environmental Research to assess the impact of PLT's pre-K-8 curriculum. PLT's curriculum was shown to work in all classes when teachers implement the activities as intended. PLT effectively increases both environmental knowledge and concern about environmental quality.

PLT is balanced on value-sensitive topics. The activities and materials are designed to treat issues fairly and do not advocate any particular point of view. PLT recognizes that people need information from a variety of sources in order to make their own informed decisions. In short, PLT helps teach students how to think about complex environmental issues, not what to think. For more information about PLT, check out their website at www. plt.org or contact their national office at (202) 463-2462. The address is 1111 19th Street, NW, Suite 780, Washington, DC 20036.

A number of organizations have used PLT resources in their own programs. Some of these reports are available in the databases of Educational Resources Information Center (ERIC) and Eisenhower National Clearinghouse (ENC) collections. To read about these resources and learn where to get them, search the ERIC or ENC collections online or at a local library or university.

Resource from ERIC

Iozzi, L., and Halsey, B., Jr. *Environmental Education Activity Guide: Pre K-8.* Washington, D.C.: American Forest Foundation, 1993. ED405172.
     Project Learning Tree's activity guide is arranged under five major themes: Diversity, Interrelationships, Systems, Structure and Scale, and Patterns of Change. Each theme covers the areas of environment, resource management and technology, and society and culture.

Resource from ENC

Exploring Environmental Issues: Focus on Forests. Washington, D.C.: American Forest Foundation, 1995. ENC-006314.
     This module, developed by Project Learning Tree, uses forest issues as a focus for students in grades 9-12 to investigate and define an environmental issue, identify key players and points of view, generate alternative solutions, recognize and weigh trade-offs, listen to their peers, and make personal and group decisions.

Project Learning Tree Independent Study and Evaluation. Seattle: University of Washington, 1977. ENC-011274.
     Prepared by the Bureau of School Service and Research and submitted to the American Forest Institute and the Western Regional

Environmental Education Council, this is an evaluation of the Project Learning Tree educational materials for the elementary, intermediate, and secondary levels.

Acknowledgement: The authors thank Kathy McGlauflin and Tess Erb of Project Learning Tree.

Environmental Education and the Social Studies: Making the Connections

by Michele Archie, Bora Simmons, Joe E. Heimlich, and Sabiha S. Daudi

In the 30 years since environmental education emerged as a field, many social studies teachers have found themselves caught between believing in the importance of environmental education and feeling pressed to find the time to fit it in. Nationwide polls tell us that 96% of parents support teaching environmental education in the schools. When teachers are asked if environmental education is important, an overwhelming majority (97%) agrees. Yet in practice, few teachers consistently include environmental education in their curriculum.

One reason for this gap is that environmental education often is seen as an add-on, as something that only science teachers do, or as an activity used to celebrate Earth Day or Arbor Day. In reality, environmental education offers social studies teachers and students a powerful tool for integrating what they're learning in a real-world context. At its heart, environmental education is an integrative undertaking. Learning about the environment and environmental issues is a continuing lesson in interconnectedness. Environmental education offers opportunities for teaching across the social studies curriculum, integrating methods and ideas from history, civics, geography, and economics to help students develop the skills they need to understand connections in the environment.

One needs to go no further than the national standards to find ample evidence of the connections among the social studies and environmental education. Taken singly, each set of national standards in social studies (such as history, social studies, civics, economics, geography) incorporates learning and instructional goals that are aligned with those of environmental education. In *Geography for Life: The National Geography Standards*, for example, one finds, "how physical and human processes together shape places" and "ways in which different people perceive places and regions." Similarly, essential components of environmental education are found throughout the *Curriculum Standards for Social Studies*. The performance expectations include, "consider existing uses and propose and evaluate alternative uses of resources and land in home, school, community, the region and beyond," and, "apply knowledge of economic concepts in developing a response to a current local economic issue such as how to reduce the flow of trash into a rapidly filling landfill."

Environmental education programs can be used to teach a variety of social studies concepts. As your students learn about the difference between human wants and needs or how different cultures interact with their environment, environmental education curriculum materials supply educationally sound lesson plans. As an example, the curriculum unit, *What Do We Need to Live on Planet Earth? A Case Study of the Traditional Rural Life in East Africa*, developed for grades 2-4 by the Stanford Program on International and Cross-Cultural Education (SPICE), provides a series of activities specifically designed for use in social studies and history. According to its developers, the unit "may be used at any point in the curriculum where students are asked to learn about people from other cultures, and to examine contemporary or more traditional lifestyles."

While studying specifically about the Masai and Kikuyu, students explore the basic question of what humans and other animals need to survive. As part of the unit, students are asked to build dwellings, considering both their form and function; examine the habitat needs of elephants and how these relate to the needs of the Masai; and make predictions about the use of water. Units such as this go a long way to providing examples of how to meet the goals of social studies and environmental education.

Creating a comprehensive, cohesive environmental education program that meshes with the standards-based social studies curriculum can be a daunting task. There are several resources now available to help teachers integrate environmental learning into social studies programs. *Excellence in Environmental Education: Guidelines for Learning (K-12)* sets expectations for performance and achievement in fourth, eighth, and 12th grades and suggests a framework for effective and comprehensive environmental education programs and curricula. These guidelines are also linked to national disciplinary standards. They demonstrate how environmental education can be used to meet standards set by the social studies.

Social studies teachers can find more help in *The Environmental Education Collection: A Review of Resources for Educators* (Vols. 1, 2, and 3), published by the North American Association for Environmental Education (NAAEE). These collections are designed to help teachers select instructional materials that meet their needs and the needs of their students. Each volume reviews a broad range of K-12 educational materials (for example, curriculum guides, videos, and CDs). All of these materials have gone through an extensive evaluation by teachers, environmental educators, and content specialists.

The Biodiversity Collection: A Review of Biodiversity Resources for Educators, published by World Wildlife Fund in association with NAAEE, is another resource. This compendium of exemplary K-12 resources is designed to help educators find high-quality biodiversity education materials, including curriculum guides, children's books, posters, and multimedia resources. All of the resources are indexed by grade level, subject area (for example, social studies, economics), and topic (for example, energy, population, solid waste, and social action).

References

Roper Starch Worldwide. *The National Report Card on Environmental Knowledge, Attitudes and Behaviors.* Washington, D.C.: National Environmental Education and Training Foundation, 1997.

World Wildlife Fund. *Windows on the Wild: Results of a National Biodiversity Education Survey.* Washington, D.C., 1994.

Geography Education Standards Project. *Geography for Life: The National Geography Standards.* Washington, D.C.: National Geographic Research and Exploration, 1994.

National Council for the Social Studies. *Expectations of Excellence: Curriculum Standards for Social Studies.* Washington, D.C., 1994.

Murphy, Carol, and Wallace, Kendra R. *What Do We Need to Live on Planet Earth? A Case Study of the Traditional Rural Life in East Africa.* Stanford, Calif.: Stanford Program on International and Cross-Cultural Education.

North American Association for Environmental Education. *Excellence in Environmental Education: Guidelines for Learning (K-12).* Rock Spring, Ga., 1999.

North American Association for Environmental Education, *The Environmental Education Collection: A Review of Resources for Educators.* Vols. 1, 2, and 3. Rock Spring, Ga., 1997, 1998, 1998.

Linking Environmental Education and English Language Development

by Kay Antunez, Joe E. Heimlich, and Sabiha S. Daudi

Researchers of second language development have accumulated extensive data on how people acquire a second language. One of the key factors in this process is the nature of the educational conditions second language students experience. It is this key factor that can unite curricular efforts of environmental education and English language development (ELD) and make possible a climate that is mutually supportive of the education goals of each.

Contributing to the creation of the conditions that best link environmental education and ELD is the process of supplying "comprehensible input." Comprehensible input is comparable to the many ways in which environmental education supports students in the construction of meaning from what is being communicated in a lesson. Based on research that suggests language development occurs best when the conditions are most favorable, ELD strives to create conditions that promote an abundance of content-based, comprehensible input for a student's acquisition of social and academic languages. English language learners (ELLs) who receive quality curricular-based ELD instruction and have grade-level literacy in their first language generally take less time to attain proficiency than those who experience learning conditions lacking in one or both of these factors. (See for example Freeman and Freeman 1995; Cummins 1996; Krashen 1982; Thomas and Collier 1997).

Environmental Education and Language Proficiency

Language proficiency refers to two distinct, interrelated kinds of language usage. Called "conversational" and "academic" language, each usage exhibits unique linguistic characteristics that demand different levels of cognitive involvement and contextual support.

Sometimes referred to as basic interpersonal conversation skills or social proficiency, *conversational language* is the easier of the two proficiencies to acquire. Much is "picked up" during concrete and purposeful communication situations that occur within the context of everyday living. Playing games, going shopping, media events, routine interpersonal exchanges, and other similar daily activities can be sources for conversational language. Fluency in conversational English, however, does not necessarily mean the ELL is ready or capable of learning academic curriculum in only English.

Conversational language mostly occurs in conditions that are context embedded and cognitively reduced. Context embedded refers to language use stimulated by and corresponding with what is happening in the immediate surroundings. Comprehension is supported through the nonverbal cues that accompany language, reducing the need to understand the individual words or ideas, which is language that is cognitively reduced. Environmental education, by its design to engage learners in the physical reality of their immediate environment, has the potential to ground learners in the conversational language based on experience.

Academic language proficiency is directly related to cognitive development and school achievement. It requires complex thinking and literacy skills, the capacity to process abstract and cognitively demanding content, and the ability to respond to questions or draw conclusions based on the "logic" of instruction or textbooks, rather than from immediate or personal experiences. The ability of an educator to move students from experience to abstract thinking can be enhanced by connecting learners with the concepts through physical or affective engagement, as is often done in environmental education.

Environmental Education as a Means for Providing "Comprehensible Input"

Comprehensible input is used to describe what educators can do to create instructional conditions that resemble a conversational setting for the purpose of developing both academic and conversational language proficiency (Krashen 1982). The emphasis on the development of academic English distinguishes ELD from more traditional English as Second Language (ESL) programs.

Environmental education's constructivist orientation, which is active, hands-on, and student-centered, embodies instructional conditions highly compatible with the needs of ELLs. By using the academic language and content from various disciplines while requiring the use of different cognitive processes, environmental education lessons generally provide a certain amount of comprehensible input, therefore providing a comfortable learning context for students still in the process of acquiring English as a second language.

However, since the acquisition of a second language is a developmental process that usually takes from five to seven years to complete, the amount of comprehensible input a lesson is capable of providing depends on the developmental stage of an ELL. Simply put: The more advanced students are in the process of ELD, the more likely it is that lessons are supplying the kind of comprehensible input required for learning academic content.

References

Cummins, J. *Negotiating Identities: Education for Empowerment in a Diverse Society*. Los Angeles: California Association for Bilingual Education, 1996.

Freeman, David, and Freeman, Yvonne. "SDAIE and ELD in Whole Language." *California Association for Bilingual Education Newsletter* 18, no. 4 (1995).

Krashen, S. *Principles and Practice in Second Language.* New York: Pergamon, 1982.

Thomas, W.P., and Collier, V.P. *School Effectiveness for Language Minority Students.* Washington, D.C.: National Clearinghouse for Bilingual Education, 1997.

Resources from ERIC

Thomas, W.P., and Collier, V.P. "Two Languages Are Better than One." *Educational Leadership* 55 (December-January 1997-1998): 23-26. EJ556857.

This article suggests that instead of viewing English learners as a problem needing remediation, educators should build an enriched bilingual program for all students. In successful two-way programs, both language groups stay together throughout the school day, serving as peer tutors for each other.

Whitelaw-Hill, Patricia. "Setting Achievement Goals for Language Minority Students." *News from READ* (Winter 1995): 1-2, 4. ED391353.

This article by Patricia Whitelaw-Hill, executive director of READ, addresses the issue of the education of limited-English students in public education.

Ensuring Comprehensible Input for English Language Learners: Strategies for Environmental Education Providers

by Kay Antunez, Joe E. Heimlich, and Sabiha S. Daudi

Using a variety of strategies when communicating with English language learners (ELLs) is one of the best ways environmental educators can help make lessons more comprehensible. There are three key lesson assessments and adjustments that are usually necessary: 1) the type of instructional language used to carry out the lesson, 2) recommended instructional practices, and 3) the lesson structure.

It is recommended that the educator consider how many of the recommended strategies suggested below already are incorporated into an environmental education lesson or activity and then modify the activity appropriately. Adjusting environmental education lessons to attain the best "fit" between the three key areas mentioned above and the linguistic needs of ELLs entails a process of customizing the basic teaching strategies in order for the lesson to be compatible with ELLs' English language development. Ideas in this article draw primarily on the work of Jim Cummins (1996), David and Yvonne Freeman (1995), Stephen Krashen (1981, 1991), and Alfredo Schifini (1994).

Assessing and Adjusting the Instructional Language

When speaking with or providing reading material to ELLs, it is important to consider the level of complexity of the language used by the educator and the situational support included in the learning activities. In the beginning of a lesson, ELLs need comprehensible input within their linguistic comfort zone; as a lesson unfolds, they need to begin experiencing input (words, concepts, structure) slightly beyond what they themselves can produce. This is referred to as "input + one" or language that is within the learner's linguistic stretch zone. The more cognitively demanding language becomes, the more necessary are contextual aids for negotiating the meaning of the language used. Some of these aids include:

Using contextual clues: Use nonverbal, visual, and tactile materials that correspond with the language being used. Act out words, make drawings, use graphic organizers, and write down key words and ideas as they are said;

Modifying the linguistic output: Accentuate key vocabulary, emphasize and pause where needed, pause between key ideas, restate (paraphrase) in different ways and regulate sentence length, use cognates (words similar across languages) wherever possible, introduce increasingly sophisticated vocabulary for the same concept (for example, "accelerate" for "faster," "discuss" for "talk about"), control use of idiomatic expressions.

Checking frequently for understanding: Use a range of open-ended questions to promote active listening, generate extended responses, and invoke higher-level thinking skills. Ask questions as the lesson unfolds to identify concepts and vocabulary that may need to be retaught.

Assessing and Adjusting Instructional Practices

Three broad strategies can be used to make a lesson's content more accessible to ELLs. The first is to *present the lesson with ELLs in mind.* Begin lessons with concrete and experiential activities and move to the abstract and conceptual. Move from cognitively less to more demanding language and tasks. Avoid dividing concepts into parts and losing the "whole." Repeat lessons, especially for ELLs in the beginning stages. Use a variety of learning modalities, and focus less on the communication accuracy and more on the effort to communicate. Have listening and speaking activities precede reading and writing activities.

The second strategy is to *encourage active language production.* This can be achieved by extensively using cooperative grouping and peer-pairing. Use both native-language groups or pairs when previewing or reviewing a lesson and mixed-language groups while constructing knowledge. Peer and cross-age tutoring, collaborative projects designed and presented by cooperative groups, integration of writing and reading into collaborative and individual projects, and "reporting out" new knowledge all help ELLs consolidate and extend knowledge.

The third strategy is to *strengthen language development and knowledge construction.* This refers to organizing lessons drawing on students' prior knowledge and background; using a theme-based approach to organizing lessons; helping learners choose themes that connect with and can improve the quality of their personal, family, and community life-conditions; and striving to connect lessons and themes to real life. Provide appropriate primary language support whenever possible.

Assessing and Adjusting the Lesson Structure

Many environmental education materials, such as Project Learning Tree, Project WILD, and Project WET, include ways of connecting students' prior knowledge to the new knowledge in the lesson, a series of instructional or procedural steps, suggestions for assessment, and ways to extend or enrich the lesson. When teaching ELLs, two additional components are needed.

Providing primary language support helps ELLs build on knowledge they already have about the content because academic knowledge in one's primary language can provide a foundation for acquiring the academic language for the same kind of knowledge in a second language. To attain this, involve family and community members, use the primary language to review lessons, use nonverbal activities (art projects, walks, drama) to introduce new ideas in mixed-language cooperative groups, and provide family homework assignments.

Connecting ELLs' prior knowledge to the new knowledge allows ELLs to make connections and inferences about the meaning of the English being used because they already hold the academic concepts regarding the subject. Other suggestions include using visual aids to stimulate observation and discussion, providing manipulative and concrete objects to stimulate language, incorporating the backgrounds of the students to elicit different cultural knowledge of the content, and using writing prompts.

Familiarity with the generalized strategies for creating comprehensible input while teaching environmental education lessons achieves three main outcomes:

First, it provides a source for assessing and adjusting the instructional language and practices to be used when teaching the main body of an environmental education lesson to a mixed group of ELLs and English-only students.

Second, it gives a sound foundation for customizing instructional language and practices to "fit" with the characteristics of specific language proficiency levels of ELLs.

Third, collectively, these general strategies create a positive social and cultural learning context where students' cultural knowledge is expressed, shared, validated, and used to help them construct new knowledge. When this occurs, ELLs are more fully motivated to invest themselves in the learning process. The sharing and constructing of collective knowledge among teacher and students is the backbone of multicultural education and is far more profound in creating cross-cultural understandings, multicultural friendships, enhanced self-esteem, and equalized status relations than is the more typical holidays, heroes, and traditions approach.

References

Cummins, J. *Negotiating Identities: Education for Empowerment in a Diverse Society.* Los Angeles: California Association for Bilingual Education, 1996.

Freeman, David, and Freeman, Yvonne. "SDAIE and ELD in Whole Language." *California Association for Bilingual Education Newsletter* 18, no. 4 (1995).

Krashen, S. *Second Language Acquisition and Second Language Learning*. London: Pergamon, 1981.

Krashen, S. *Bilingual Education: A Focus on Current Research.* Occasional Papers in Bilingual Education. Washington, D.C.: National Clearinghouse for Bilingual Education, Spring 1991.

Schifini, A. "Language, Literacy, and Content Instruction: Strategies for Teachers." In *Kids Come in All Languages: Reading Instruction for ESL Students*, edited by K. Sprangfenberg-Urbschat and R. Pritchard. Newark, N.J.: International Reading Association, 1994.

Capacity Building for Environmental Education: How Does It Help Teachers?

by Kerry Eastman, Dan Sivek, Sabiha S. Daudi, and Joe E. Heimlich

Envision a future where preservice teachers receive quality training in environmental education, school administrators understand and support environmental education, and funding is available for special projects and field trips that bridge classroom activities and the community. These are some of the goals of environmental education capacity building.

Environmental education capacity-building efforts strengthen organizations and individuals who are working toward the implementation and maintenance of comprehensive environmental education programs at the state and local level. Environmental education capacity-building efforts target leaders at the state and local levels because these leaders have the ability to act to achieve comprehensive environmental education programs.

A comprehensive environmental education program comprises several components within three different categories: Structure, Program, and Funding. Examples of state-level components include environmental education grants programs, a state environmental education advisory board, and both preservice and inservice teacher training. Local components could include an environmental education coordinator, locally developed curricula, and local grants or donations.

Benefits to Teachers

Teachers receive both direct and indirect benefits from many of the programs and services resulting from environmental education capacity-building efforts. Teachers benefit directly from increased access to environmental education resources and other environmental educators through websites, such as EdGateway and EE-Link, state and regional environmental education organizations, and the North American Association for Environmental Education (NAAEE). Through the support of the NAAEE, the *Environmental Education Materials: Guidelines for Excellence*, *Excellence in Environmental Education: Guidelines for Learning (K-12)*, and *Excellence in Environmental Education: Guidelines for Learning (K-12) Executive Summary and Self Assessment Tool* are available to teachers. These can be used to evaluate materials used in the classroom to teach environmental education, and they illustrate that environmental education can be a tool for meeting state and national subject area standards for learner achievement.

Most state environmental education associations offer annual conferences enabling teachers to pursue professional development opportunities in environmental education. Many state environmental education associations are developing and offering leadership clinics modeled after the Annual Leadership Clinic offered by North American Association for Environmental Education (NAAEE) and the National Environmental Education Advancement Project (NEEAP). The leadership clinics offer workshops and training specifically to increase the leadership skills of individuals and organizations. Many teachers, working with their state environmental education association, attend the Annual Leadership Clinics.

Teachers benefit indirectly because environmental education capacity-building efforts enable more effective state environmental education organizations, which can in turn support teachers with resources, conferences, and funding for environmental education projects. Some environmental education capacity-building efforts have helped states develop grant programs to support environmental education projects at the local level. Additionally, a stronger state environmental education association builds stronger networks among environmental education practitioners within the state to provide a forum for idea and experience sharing.

Who's Involved?

In the early 1990s several environmental education leaders came to believe that the field of environmental education needed to become better organized; and specific organizations were established, with environmental education capacity building as their main goal.

The National Environmental Education Advancement Project (NEEAP) was established in 1991 with the mission to assist state and local environmental education practitioners in promoting and enhancing environmental education efforts in their state. NEEAP is a member of the Environmental Education and Training Partnership (EETAP). EETAP is a consortium of organizations managed by NAAEE and funded by the U.S. Environmental Protection Agency. The purpose of EETAP is to coordinate the efforts of its partners in order to increase the number of education professionals trained in environmental education.

NAAEE offers an annual conference, publications, and other networking programs and forums. NAAEE developed the Affiliates Partnership program in 1991 to promote and support the work of environmental educators throughout North America. The Affiliates Program is a network of professional environmental education associations throughout North America.

The State Education and Environment Roundtable (SEER) was formed to provide a forum for environmental education specialists in state education agencies. SEER's mission is to aid state education agencies in their efforts to improve learning by using the environment as an integrating concept.

EdGateway is an interactive website sponsored by the WestEd Eisenhower Regional Consortium for Mathematics and Science Education and EETAP. The website offers users the opportunity to participate in online discussions to share information with other educators, access to information about environmental education and education reform, and calendars of events and conferences.

How Can Educators Benefit?

By combining the efforts of individuals and organizations working toward building comprehensive environmental education programs, it is possible to create the synergy needed to develop an environmentally literate citizenry. Listed are some suggestions for getting involved in, and staying informed about, environmental education capacity-building efforts.

- Join your state environmental education association. State associations rely on their members for support in order to be able to provide individuals with the resources for environmental education. By joining your state environmental education association, you will be helping your state toward stronger environmental education programs. Contact NAAEE to locate the contact for your state association.
- Join the North American Association for Environmental Education. NAAEE can provide you with resources, a national newsletter, and networking among other environmental education providers (http://www.naaee.org).
- Contact such organizations as the National Environmental Education Advancement Project (NEEAP, http://neeap. uwsp.edu) and NAAEE for organizational and leadership training and support. The goal of these organizations is to provide these services specifically for the environmental education field.
- Visit EE-Link (http://eelink.net) and EdGateway (http:// www.edgateway.net) to find out about resources and opportunities in the field of environmental education. EdGateway also has interactive discussion sites to provide you with a

chance to communicate and share ideas with others in the environmental education field.

References

State Capacity Building Commission. "State-Level Capacity Building for EE: Next Steps." Concept Paper. NAAEE, 1998. http://neeap.uwsp.edu

Ruskey, Abby, and Wilke, Richard. *Promoting Environmental Education: An Action Handbook for Strengthening EE in Your State and Community*. League City, Texas: National Association of Conservation Districts, 1994. ED377076.

Ruskey, Abby. "Toward De-Mystifying 'EE Capacity Building'." *RockEE News* 1, no. 2 (Summer 1999).

Resources from ERIC

Mazrek, Rick, ed. *Pathways to Partnerships: Coalitions for Environmental Education*. Selected Papers from the 22nd Annual Conference of the North American Association for Environmental Education, Big Sky, Mont., 1993. ED406175.

Simmons, Deborah, et al. *Environmental Education Materials: Guidelines for Excellence*. Troy, Ohio: North American Association for Environmental Education, 1996. ED403145.

The Environmental Education Collection: A Review of Resources for Educators. Vol. 1. Troy, Ohio: North American Association for Environmental Education, 1997. ED416079.

Office of Environmental Education. *Resource Materials to Support Your Environmental Education Efforts*. EE-TIPS (Environmental Education Technical Information Packages). Washington, D.C.: Environmental Protection Agency, 1996. ED419681.

Environmental Education Beyond Nature Study

The social, cultural, economic, and political considerations in environmental education move the field far from the study of nature. Much work in environmental education is done around individual actions and the relationships of those behaviors to the larger environment, the "act locally, think globally" concept. The decisions people make — whether for work, recreation, nutrition, or protection — affect the environment both directly and in a multitude of indirect ways. What may seem like a simple decision — for example, to buy some gum — has ramifications on environmental systems. The production, packaging, transportation, marketing, and consumption of that package of gum affects resources, labor, economic systems, air quality, water quality, energy, and a host of other issues.

Of course, one pack of gum does not make or break the environment. Rather, all the decisions people make have a cumulative effect. But environmental education is focused on internalizing the locus of control for the learner and letting them discover how their decisions affect their immediate world and the larger world.

Environmental education is not just about nature study. It is about the Earth and the systems of the Earth. Therefore some of the current themes in environmental education are decision making and decision-making theory. Helping learners understand how they make decisions and learning how to be informed (and continually more informed) in their choices is an important component of environmental education. And because environmental actions take place within a social context, another important issue is that of equity and environmental justice.

This section provides a basis for understanding both the theories and the applications of decision making and justice within the context of applied environmental education.

Environmental Decision Making: A Basis for Informed Choices

by Sabiha S. Daudi and Joe E. Heimlich

The environment in which we conduct our daily tasks includes both a social and a physical component. We perceive our natural environment through a perpetual screen that consists of a woven pattern of past experiences and cultural influences. Our life experiences, when they interact with our natural environment, can lead to a better understanding of environmental and related issues, improved management of natural resources, and concerns about their degradation.

For making informed choices about environmental issues that affect our daily life, we need to be aware of the issue, all related problems, and possible alternatives in order to make best decisions about our environmental concerns.

Environmental Decision-Making Process

Decision making has been defined by scholars as "the act of choosing [a course of action] among a set of alternatives under conditions that necessitate choice." If this definition is examined carefully, it will be clear that, at some point in our lives, we have had to make choices that have affected the natural environment directly or indirectly. In certain situations — littering on the streets, for example — we have felt compelled to make a choice and have made one, even if it was to do nothing. Such environmental decision making affects our lives in the short term and the health of our planet in the long run.

In order to allow this decision-making process to be meaningful and functional in the context of the natural environment, we need to carefully study the alternatives available to us; understand the immediate, as well as long-term, effects; reduce uncertainty about the outcome; choose a course of action; and implement the chosen plan while continuously reflecting on and monitoring progress.

The decision-making process plays an important role in developing strategies for dealing with environmental issues that are of concern to citizens. These processes become even more important when the outcome relates to matters of human survival. In such cases, environmental decisions usually are made on the basis of short-term outcomes to deal with the effects. However, to achieve the goals of *sustainable development*, defined as "improving the quality of human life while living within the carrying capacity of supporting ecosystems," we need to plan to leave the environment of planet Earth for future generations at least as we found it. It therefore becomes essential that everyone — corporations, communities, and individuals — be involved in planning progressive strategies at all stages of environmental decision making.

Environmental educators can focus on developing the learners' decision-making skills by providing experiences that explore solutions and alternatives to environmental concerns. This could enhance learners' ability to make informed choices when faced with alternatives that have long-term, as well short-term, effects on the natural environment.

An example of such an effort is the *Environmental Issues Forum* (EIF), a program of the North American Association for Environmental Education (NAAEE) that helps adults in their communities and high school and college students discuss controversial environmental issues. The EIF program brings the creativity and energy of these informal discussions into a more formal setting.

There are a number of other educational resources available for educators that focus on the environmental decision-making process and its application to the management of natural resources. These documents also explore case studies and situations where environmental decision making has played a visible role in making informed choices.

Resources from ERIC

Rowland, P. McD., and Adkins, C.R. "Developing Environmental Decision-Making in Middle Schools." Paper presented at the meeting of the World Congress for Education and Communication on Environment and Development, Toronto, 18 October 1992. ED361168.

This paper presents Rowland's Way of Knowing and Decision-Making Model for curriculum development and how it can be applied to environmental education curricula. An example is provided from a seventh-grade class that investigated the issues surrounding the Glen Canyon Dam controversy through the study of native and exotic plants and endangered fish and their habitat.

McConney, A.W., et al. "The Effects of an Interdisciplinary Curriculum Unit on the Environmental Decision-Making of Secondary School Students." Paper presented at the 67th annual meeting of the National Association for Research in Science Teaching, Anaheim, Calif., 26-29 March 1994. ED368566.

This two-phased research study focused on development of an interdisciplinary unit on sustainable development in tropical rainforests by simulating an environmental problem. The students were encouraged to develop and decide on a solution, having weighed a spectrum of possibilities previously explored in class activities and discussions.

Chociolko, C. "The Experts Disagree: A Simple Matter of Facts Versus Values?" *Alternatives* 21 (July-August 1995): 18-25. EJ507438.

This article challenges the assumption that disagreement among environmental experts is a rare occurrence and suggests that the public offers valuable expertise that should be included in the environmental decision-making process. It also looks at forms of bias and methods for improving the quality of technical knowledge and expertise. Sidebars discuss actual conflicts.

Cohen, B.L. "The Role of Darwinism in Environmental Decision Making." *Bulletin of Science, Technology and Society* 10, nos. 5-6 (1990): 270-74. EJ421885.

Inconsistent decision making of the United States government on environmental issues is described in this article. Also discussed are the health risks of the alternatives, environmental group leadership, the role of news media, and the role of government.

Resources from ENC

Helpern, N.A. *Kids and the Environment.* Software. Mass.: Tom Snyder Productions, 1994. ENC-005134.

This software program focuses on environmental issues and presents children with a realistic dilemma filled with difficult choices. The dilemma is the garbage strewn on the soccer field of a school. The students, as captains of the soccer team, must determine how to use the field for practice by discussing long- and short-term solutions, making choices, and evaluating the consequences of their choices. They also must weigh alternatives, initiate change, set priorities, and establish goals based on those priorities.

Heil, D.; Allen, M.; Cooney, T.; Matamoros, A.L.; Perry, M.; and Slesnik, I. *Wetlands: Making Decisions.* Kits. Glenview, Ill.: Scott Foresman, 1996. ENC-004031.

This module, designed for grade 6, allows the students to address the issue of today's vanishing wetlands. Students also learn how understanding the interactions of natural systems, such as wetlands or the food chain, enables people to make appropriate, informed, environmental decisions.

Lenk, C. *Rainforest Researchers.* CD-ROM kit. Mass.: Tom Snyder Productions, 1996. ENC-006492.

This CD-ROM kit for students in grades five to eight is intended to involve them, through scientific analysis, decision making, and teamwork, in botanical and ecological research in the rainforests of Indonesia. The program also builds and reinforces critical-thinking and problem-solving skills.

Wrigh, R.G., and Moyer, A.W. *Oil Spill!* Kits. Boston: Innovative Learning Publications, Addison-Wesley, 1995. ENC-007878.

This unit, including a teacher's guide and videotapes, is one of a series of Event Based Science Modules that engage students by having them watch television news coverage of a real event from the earth sciences and read about it in the newspaper reports.

Decision Making:
An Important Process for
Environmental Educators

by Joe E. Heimlich and Sabiha S. Daudi

If indeed one of the goals of environmental education is to create citizens capable of making sound decisions regarding the environment, then it is important that environmental educators understand not only environmental issues, but also the decision-making process.

Decision making has been discussed and studied since the Greeks wrote on discourse and rhetoric. Within the body of writing on decision making, there is much presented on persuasion and how to encourage understanding of issues. From this literature, many concepts were reiterated and revised through the centuries; and concepts of decision making appear in literature, drama, poetry, and history.

John Dewey was instrumental in bringing the concepts of decision-making theory into education. He presented what he saw as the "Reflective Thinking Process," which involved four phases: the problem phase, the criteria phase, the solution phase, and the implementation phase. Dewey believed that good learning was based on the ability of the learner to discern problems, identify what would satisfy the problem, consider alternatives for meeting the criteria, and then implement the decision.

Most of the research in decision making is not within education, however. Psychology is the discipline within which most research and discussion on decision making is based. It is applications of theory that abound in leadership, business, philosophy, communications, and education.

From the literature, there are subtle but important distinctions among words that often are used less carefully as synonyms. A *conclusion* relates directly to facts but not to actions, while a *judgment* relates directly to values, rather than facts or actions. A *decision* relates directly to proposed actions, which is why it is often used as an outcome goal of environmental education. The purpose of decision making, whether by an individual or a group, is to optimize the outcomes of an action.

There are four assumptions in decision-making theory:

1. In making a decision, an individual always uses a set of rules that relates to their experience, understanding of the issue and effects, and their social/cultural background.
2. The context in which the decision must be made provides the "specificity" of the decision.
3. The generality of the decision, or the transferability of the action to other situations, affects an individual's decision.
4. The strength of the heuristics (or proofs) or the individual's perception of the strength of the heuristics will affect the ultimate decision.

These assumptions suggest four important educational foci for environmental decision making. It is possible to work with learners on the "rules" they use for decision making to improve the process. Likewise, it is possible to examine the context in which the decision must be made to understand how situations affect decisions. The compilation or transfer of a decision to another context is sound constructivist theory, and understanding the heuristics is basic training in logic and reasoning.

As educators, we can reconsider how we teach environmental decision making by asking ourselves:

- Are we after a decision, a conclusion, or a judgment?
- If we are seeking a decision by the individual, do we understand their rules for making decisions and should we make these more explicit or implicit?
- How specific is the decision we are after? What are the transferable elements of the decision?
- How sound is the reasoning used in making the ultimate judgment?

As environmental educators, our challenge is to transfer and apply this theoretical base for decision making to the learners' life experiences, thus making them more relevant and longer lasting.

There are a number of resources available on the databases of the Educational Resources Information Center (ERIC) and Eisenhower National Clearinghouse (ENC) collections.

Resources from ERIC

Haakonsen, H.O., et al. *A Self-Instructional Approach to Environmental Decision-Making: Focus on Land Use*. New Haven, Conn.: Environmental Education Center, Area Cooperative Educational Services, 1976. ED133206.

This paper provides an overview to the Land Use Decision Making Kit. The Land Use kit includes 16 audio-tutorial units and a variety of supplementary materials. Each audio-tutorial unit consists of a program mix of cassette tapes, guide sheets, visuals, pamphlets, and issue-keyed problems.

Hounshell, P.B., and Coble, C.R. "Environmental Decision Making in the Classroom." *Science Teacher* 40 (April 1973): 21-23. EJ074473.

Relevance creates a dilemma for the teacher because solutions to problems are controversial in contemporary society. This paper argues that students still need to explore these problems. It is hoped that they can come to understand and appreciate the situations and comprehend the alternatives.

McConney, A.W., et al. "The Effects of an Interdisciplinary Curriculum Unit on the Environmental Decision-Making of Secondary School Students." Paper presented at the 67th annual meeting of the National Association for Research in Science Teaching, Anaheim, Calif., 26-29 March 1994. ED368566.

The centerpiece of this interdisciplinary unit is the investigation of a simulated environmental problem that requires students to develop and then decide on a solution, having weighed a spectrum of possibilities previously explored in class activities and discussions.

Resources from ENC

Florida State University. *Decisions About Science*. Newton, Mass.: WEEA, 1983. ENC-002530.

This program, developed for middle school students, focuses on expanding each student's self-concept and increasing each student's decision-making capabilities. One section deals with information about social, natural, and human-made environments and the conflicts between them. That section is followed by opportunities for students to make environmental decisions.

United States Forest Service. *Investigating Your Environment*. Ogden, Utah: U.S. Forest Service, Intermountain Region, 1994. ENC-002687.

The complete edition of *Investigating Your Environment*, designed to assist teachers in expanding their repertoire of teaching activities that focus on environmental issues, includes descriptions of how to investigate, interpret, and make decisions about the environment, land use, and natural resources.

Spence, M.S.; Fix, M.A.; Shafer, M.C.; and Browne, J. *Statistics and the Environment*. Chicago: Encyclopedia Britannica Educational Corporation, 1998. ENC-011851.

This student text and accompanying teacher guide for grades 7 and 8, part of the Mathematics in Context series, contains an interdisciplinary unit that emphasizes statistical reasoning in making plans to convert a fictitious island into a nature reserve.

Decision Making: Grounded in Behaviors

by Joe E. Heimlich and Sabiha S. Daudi

Because decisions relate directly to actions, it is logical that much of the study of decision making grows out of behaviorist theory and study. Some of the concepts from this understanding of decision making are outlined here so that environmental educators can better appreciate the solid grounding of the work we do.

Impulse behaviors are those things we do without reflection or consideration. Impulse behaviors may be reflexive or intuitive behaviors; but they are not the result of learning, training, or practice. Jumping when startled is an impulse.

Routine behaviors are those that are based on familiar situations. Routines may be isolated habits or habits that are ingrained in a specific situation or location. The series of things we do when getting up in the morning and most of the decisions we make when driving are routine behaviors.

Casuistic behaviors are based on accepted norms or mores. Usually casuistic behaviors are culturally defined as "right and wrong" behaviors. But these behaviors also can be based on actions modeled by cultural models or heroes. Standing in line is a classic casuistic behavior.

Thoughtful behaviors are those based on reflection and intentionally chosen by an individual.

In environmental education, we often try to teach behaviors as "thoughtful" actions and hope that they are accepted by the individual. Ultimately, we desire the behaviors to become parts of "routine behaviors" or habitual actions by an individual.

Yet in our teaching for behavioral decisions, there are often gaps that are considered by behaviorists to be "structural" in nature. Some of the theorists define these structural problems as an issue of discursive consciousness versus practical consciousness. Discursive consciousness is a label given to behaviors that can be expressed in language. The problem arises with the practical consciousness actions, which are those based on tacit knowledge and bound with action. Often, individuals cannot put into words an adequate argument for why they do or should do a particular behavior, as is often the case with "environmentally appropriate" behaviors. We do them "just because." The inability to isolate the action in language can make the discussion of the action appear simplistic or superficial or, at best, make it sound like a justification for the action that is not scientifically sound.

Larson suggested a "single question form" for decision making that would help move practical consciousness to a discursive level. First, we must ask "what is the single question, the answer to which accomplished the purpose" or the decision. Second, we begin to ask what subquestions must be answered before we can answer the first question. We then seek sufficient information to answer these subquestions confidently. After we have reasonable answers to the subquestions, we can, assuming correctness, identify the best solution to the problem. In moving to this form of decision making, we are not asking the individual to discuss behaviors for which they have no language or understanding. Rather, the dialogue has moved into the consequence model of decision making, which asks that a decision be based on careful consideration of the alternative outcomes for the decision.

Many environmental educators use mind-maps or decision-trees, which are variations on the consequence model. While these tools are valuable for exploring the alternatives and consequences, it is important that the environmental educator using these tools understand that underlying the use of consequences are assumptions that individuals have preferences, that the preference is transitive, and that there is independence in the decision.

Resources from ERIC

McLagan, P.A. *Behavior Theory and Adult Education.* Paper presented at the Adult Education Research Conference, St. Louis, Mo., April 1975. ED110851.

   The paper proposes that when behavior change is a major target of an adult education program, the designer must consider the basic targets for behavior change efforts, that is, behavior goals and plans, basic knowledge and needed skills, physical environment, and reinforcers of behavior.

Presbie, R.J., and Brown, P.L. *Behavior Modification: What Research Says to the Teacher.* Washington, D.C.: National Education Association, 1976. ED118563.

   This report reviews findings from the extensive research on behavior modification. It summarizes the more important, practical, concrete, and classroom-tested procedures that are shown to be effective in improving students' academic and social behavior.

Teaching About Ethics and the Environment. Cocoa and Cape Canaveral: Brevard County School Board, Florida Solar Energy Center, 1984. ED244840.

   This unit consists of activities designed to develop value systems related to the interactions of humans and their environment. The overall objectives are to teach students to evaluate their actions within an environmental context, make rational decisions in resolving environmental problems, and function in a democratic society by reaching consensus on environmental issues.

Resources from ENC

Cozic, C.P., ed. *Global Resources: Opposing Viewpoints.* Opposing Viewpoints Series. San Diego: Greenhaven Press, 1998. ENC-011905.

   This book, published for students in grades nine and up as part of the Opposing Viewpoints Series, is an anthology of essays that examine the availability of and dependence on the world's natural resources. It addresses the debate over the effects of population growth on the availability of such global resources as oil reserves, food supplies, and the rainforest.

Middleton, J.A.; Burrill, G.; and Simon, A.N. *Decision Making*. Chicago: Encyclopedia Britannica Educational Corporation, 1998. ENC-012985.

This student book with teacher guide for grades seven or eight, part of the Mathematics in Context series, contains an 11-day unit on representing data with graphs, working with the notions of constraints, and graphing inequalities and discrete functions. Decision-making skills are sharpened by addressing specific problems.

Netzley, P.D. *Issues in the Environment*. Contemporary Issues Series. San Diego: Lucent Books, 1998. ENC-012844.

This book, written for grades six to nine as part of the Contemporary Issues series, explores four issues in environmental science: the Endangered Species Act, federal park management, solid waste disposal, and the economic costs of environmentalism.

Passe, J. *School, Family, and Community Partnership: Your Handbook for Action*. Thousand Oaks, Calif.: Corwin Press, 1997. ENC-012227.

This resource book is written for anyone interested in creating a successful partnership between schools and the family and the community and is designed as a guide to the process of planning, decision making, implementing, and maintaining a successful partnership.

Decision Making:
A Hierarchy of Behaviors
for Change

by Joe E. Heimlich and Sabiha S. Daudi

Research clearly shows that there are relationships among attitudes, behaviors, and knowledge. This same research is explicit in acknowledging that, though related, there is no direct hierarchy or causal link among these variables. Such research suggests that knowledge itself does not change attitudes or behavior. Nor does changing behavior necessarily alter attitudes or knowledge.

Given such a premise, changing decisions regarding behaviors requires an understanding by the educator of behavior and how to appeal to an individual to consider action. Psychology, and especially the behaviorist and neo-behaviorist schools of thought, provide a structure for understanding the types of behaviors in which people engage and ways for intervening to help learners improve their decision making related to their behaviors. In the case of environmental education, these would be environmentally responsible behaviors, or behaviors that are conscious changes or actions that grow out of environmental messages.

Reflexive behaviors are those behaviors over which we have no (or very limited) conscious control. Things such as breathing, heart-beat, and muscle movement are at the subcortal — or below thinking — level of the brain. Think of the hammer on the joint a physician uses to test "reflexes." Some people can consciously alter some of these behaviors, but the natural process is one that precedes thinking, which explains why a person can survive after being declared "brain dead."

Preconscious behaviors refer to intuitive behaviors, which are often called reactions. These are behaviors that, if we consciously consider them, we can control but that, when we are not thinking about the behavior, lead to spontaneous reactions. If someone sneaks up behind another person and startles them, the jump, gasp, or other reaction is preconscious. Yet, if the person is aware of the "sneak attack," they may be able to not jump, gasp, or otherwise react.

Both reflexive and preconscious behaviors are not the types of behaviors we seek to address in environmental education. These are natural behaviors that are part of being a human animal.

However, where we do affect behaviors is in the *conscious behaviors* level. When learning a new behavior, we can see the behavior as an *isolated habit*. This is how we learn skills. We isolate one part of the task and learn mastery. Then we add the next part of the task. Isolated habits could be such things as consciously turning off the lights when one leaves the room or being aware of the need to separate the recyclable material from the garbage when starting to discard something. Many of the prescriptive environmental behaviors that historically have been taught are isolated behaviors, and we hope to make them habits in the learners.

Also at the conscious level is the *contextualized habit*. At the conscious level, using the same two examples from above, the contextualized habit would be that anytime the person leaves any room, they stop, turn back toward the room, turn off the lights, then leave. Or when a person is working in the kitchen, anytime they empty a container, they rinse the container and then put the container in the recycling bin. The contextualized habit is a behavior embedded in a series or pattern of other behaviors that together serve a purpose.

Contextualized habits can, over time and with consistent repetition, become *subconscious behaviors*, which are known as routines. It is when a pattern of action becomes routinized that an individual performs a series of actions or behaviors in a consistent manner without being aware of the actions. For most people, there are hundreds of thousands of routinized behaviors performed each day. Think about the patterns in the morning after getting out of bed: few people consider all the actions they undertake. In fact, many people perform their "morning rituals" without thinking about the actions at all.

At the deepest level of the subconscious, sometimes called the *post-conscious level*, are the behaviors that are done in extreme routine. These are the behaviors that cannot be changed or altered without creating cognitive dissonance. A very simplistic example may be a phone number you dial often. When asked to give someone the number, you have to "dial the number" and consciously reconsider what numbers you pushed.

For environmental education, the behaviors we hope to affect are those in the conscious and routine levels of behavior. From the literature, we can learn that we should teach, as we do, behaviors as isolated habits. But the literature suggests we go further and help the learners put the behaviors into a context, something we are less proficient at doing.

Of course, technology, knowledge, attitude, and time affect the decisions all of us make in our lives. We want to be sure that the patterns we're teaching remain at a near-conscious level so that, when learners receive new information or are challenged in their behaviors, they are able to reevaluate the behavior to make a new decision that is appropriate for their lives.

Resources from ERIC

Hounshell, P.B., and Coble, C.R. "Environmental Decision Making in the Classroom." *Science Teacher* 40 (April 1973): 21-23. EJ074473.

Relevance creates a dilemma for the teacher because solutions to problems are controversial in contemporary society. This paper argues that students still need to explore these problems and, it is hoped, they can come to understand and appreciate the situations and comprehend the alternatives.

Singer, F.S. "Cost-Benefit Analysis in Environmental Decision Making." *Journal of College Science Teaching* 7 (November 1977): 79-84. EJ196759.

Discusses how to set the ambient standards for water and air based on cost-benefit analysis. Describes marginal analysis, the basis of cost-benefit analysis, and how dynamic cost-benefit analysis is carried out with application to the automobile pollution problem.

Stamm, K.R., and Bowes, J.E. "Communication During an Environmental Decision." *Journal of Environmental Education* 3 (Spring 1972); 49-55. EJ061146.

This is an analysis of information exchange about a Corps of Engineers water-management project and the failings of current communication channels and procedures (or lack of them). Suggestions are made for increasing the contribution of communication to improved environmental decision making.

Resources from ENC

Passe, J. *School, Family, and Community Partnership: Your Handbook for Action*. Thousand Oaks, Calif.: Corwin Press, 1997. ENC-012227.

This resource book is written for anyone interested in creating a successful partnership between schools and the family and the community. It is designed as a guide to the process of planning, decision making, implementing, and maintaining a successful partnership.

Why Do We Have To? CD-ROM. Chicago: World Book, 1997. SN-010149. http://www.worldbook.com

This CD-ROM, designed for children aged three to seven, teaches lessons about getting along, making decisions, and respecting others, while presenting basic reading, matching, and decision-making skills.

Project WET. *Water, A Gift of Nature*. Las Vegas: KC Publications, 1997. ENC-013482.

This book, designed for young people in preschool through high school, teachers, and parents, provides interpretive text that illustrates subjects found in the Project WET Curriculum and Activity Guide. Project WET (Water Education for Teachers) is a national interdisciplinary water education program that is grounded in the belief that, when informed, people are more likely to participate in the decision-making process.

Role of Education in Environmentalism and Social Justice

by Sabiha S. Daudi and Joe E. Heimlich

Environmentalism in the United States grew out of the progressive conservation movement that began in 1890s. The modern environmental movement has its more contemporary roots in the civil rights and antiwar movements of the late 1960s. The more radical student activists splintered off from the civil rights and antiwar movements to form the core of the environmental movement in the early 1970s (Bullard 1990). The student environmental activists, affected by the 1970 Earth Day enthusiasm in colleges and universities across the nation, had hopes of bringing environmental reforms to the urban poor. They saw their role as environmental advocates for the poor, since the poor had not taken action on their own (Hayes 1987). They were, however, met with resistance and suspicion. Poor and minority residents saw environmentalism as a disguise for oppression and as another "elitist" movement (Schnaiberg 1980).

Morrison and Dunlap (1986) have divided environmentalism into three categories:

1. *Compositional eliticism* implies that environmentalists come from privileged class strata.
2. *Ideological eliticism* implies that environmental reforms are a subterfuge for distributing the benefits to environmentalists and costs to nonenvironmentalists.

3. *Impact eliticism* implies that environmental reforms have regressive distributional impacts.

Bullard (1990) suggests that impact eliticism has been the major sore point between environmentalists and advocates for social justice who see some reform proposals creating, exacerbating, and sustaining social inequities. Conflicts usually center largely on the "jobs versus environment" argument. Imbedded in this argument are three competing advocacy groups: 1) *environmentalists* are concerned about leisure and recreation, wildlife and wilderness preservation, resource conservation, pollution abatement, and industry regulation; 2) *social justice advocates* highlight expanded opportunity, economic mobility, and institutional discrimination; and 3) *economic boosters* have as their chief concerns maximizing profits, industrial expansion, economic stability, and deregulation.

Social justice advocates refer to the track record of environmentalists and preservationists for improving environmental quality in the nation's racially segregated inner cities and hazardous industrial workplaces and for providing housing for low-income groups. Decent and affordable housing, for example, is a top problem for inner-city residents. On the other hand, environmentalists' continued emphasis on wilderness and wildlife preservation appeal to a population that can afford leisure time and travel to these distant locations. Many wilderness areas and national parks remain inaccessible to the typical inner-city resident because of inadequate transportation. Thus physical isolation serves as a major impediment for minorities in the mainstream conservation and resource management activities.

Since the late 1970s, a grassroots social movement merging the social and environmental causes has emerged around toxic threats. Citizens are mobilized around an antiwaste theme, and the leaders of the movement have acquired new and interactive skills in areas where they have little or no prior experience.

With this merging of causes, the environmental movement has shown that it can make a difference in the quality of life we enjoy in the United States. Environmentalists and conservationists alike have a significant role in shaping the nation's development pattern when it comes to environmental impacts and land use. A dominant current challenge is for the environmental movement to embrace social justice and other minority concerns.

Formal and nonformal educational programs not only provide knowledge and information but also encourage the learner to become involved in developing a knowledge base and understanding of environmental concepts and issues. A well-developed sense of critical thinking, decision making, and problem solving leads to making informed choices in daily life.

As teachers and nonformal educators, our role thus becomes critical if we hope to inculcate understanding of civic rights and responsibilities in our learners. The capability to absorb information provided by different disciplines, the ability to seek unbiased opinions, and the desire to learn the truth are all goals of education. The same goals apply to environmental education, which focuses on environmental concerns and issues.

There are a number of teaching and learning resources available online that provide history, background, and current activities in the environmental justice movement and help educators and others understand the issues and concerns of the environmental justice movement.

References

Bullard, R.D. *Dumping in Dixie: Race, Class, and Environmental Quality*. Boulder, Colo.: Westview, 1990.

Hayes, R. "Defending the Rainforests." In *Call to Action: Handbook for Ecology, Peace and Justice*, edited by B. Erickson. San Francisco: Sierra Club Books, 1987.

Morrison, D.E., and Dunlap, R.E. "Environmentalism and Eliticism: A Conceptual and Empirical Analysis." In *Environmental Management* 10 (1986): 581-89.

Schnaiberg, A. *The Environment: From Surplus to Scarcity*. New York: Oxford University Press, 1980.

Resources from ERIC

Sarokin, D.J., and Schulkin, J. "Environmental Justice: Co-Evolution of Environmental Concerns and Social Justice." *Environmentalist* 14 (Summer 1994): 121-29. EJ496811.

This article describes the co-evolution of environmental concerns with civil rights that has found overlap in the environmental justice movement. It also discusses implications for decision making and protecting both environmental quality and civil rights.

Warren, K. "Educating for Environmental Justice." *Journal of Experiential Education* 19 (December 1996): 135-40. EJ546405.

A college teacher describes how she teaches about environmental racism using simulations and investigative trips to communities of color. Also discusses coping with paralyzing feelings, such as despair, that accompany learning about environmental racism.

Resources from ENC

Century, J.R. *Who Gets Taught, and How? Equity, Setting Standards, the Curriculum and Pedagogy*. Newton, Mass.: Education Development Center, SSI Technical Assistance, 1994. ENC-009552.

This article discusses how setting standards affects equity in education and explains that, given the governance structure of this country, states are expected to be the center of standard setting, with federal and local governments playing an important role.

Petrikin. J.P., ed. *Environmental Justice*. San Diego: Greenhaven Press, 1995. ENC-014255.

This book, a part of the At Issues series, examines the issue of environmental racism, the disproportionate number of the nation's industrial and waste facilities found in or near minority residential areas.

Maser, C. *Resolving Environmental Conflict: Towards Sustainable Community Development*. Delray Beach, Fla.: St. Lucie Press, 1996. ENC-012112.

This book advocates using conflict resolution as a way to deal with the challenges associated with protecting our natural resources and focuses on giving people the necessary philosophical underpinning for practicing a conflict resolution process, called transformative facilitation, to resolve environmental conflicts. This facilitation process aims to achieve long-term social/environmental sustainability.

Environmental Justice in a Changing World: A Historical Perspective

by Sabiha S. Daudi and Joe E. Heimlich

The call for environmental justice first arose during the 1970s with the work of such grassroots organizations as the Mothers of East Los Angeles and Chicago's People for Community Recovery. These organizations focused on addressing specific local environmental problems. The various organizations shared one unifying belief: that the poor and minorities are systematically discriminated against in the siting, regulation, and remediation of industrial and waste facilities.

Since the late 1970s, a "new" form of environmentalism has taken root in America. Among many, a grassroots social movement has emerged around the "toxics" threat; citizens are mobilized around an anti-waste theme, housing, transportation, air quality, and even economic development. Issues generally ignored by the traditional environmental agenda are brought into focus.

Focusing on the black community, Bullard suggests that "black community activists in this new movement [have] focused their attention on toxics and the notion of deprivation" (1990, p. 104). This has led to black community residents comparing their environmental quality with that of the larger society, leading to a sense of deprivation or unequal treatment. Through social and political protest, less-privileged and minority neighborhood groups aggressively challenged local development efforts they considered undesirable, becoming an effective voice for the concerns of inner-city residents. In the absence of empirical research, however, the evidence of these organizations' claims of discrimination remained largely anecdotal. As a result, their influence on policy was limited. Not until several studies appeared to substantiate their assertions did the movement gain national attention (Lambert et al. 1996).

The first major attempt to provide empirical support for environmental justice claims was conducted in the late 1970s by Robert D. Bullard, a sociologist at the University of California in Riverside. Examining population data for communities hosting landfills and incinerators in Houston, Texas, Bullard found that while black residents made up only 28% of the city's population, six of its eight incinerators and 15 of its 17 landfills were located in predominantly black neighborhoods. The presence of these facilities, Bullard suggests, not only makes black Houston the "dumping grounds for the city's household garbage," but also compounds the myriad social ills (for example, crime, poverty, drugs, and unemployment) that already plague poor, inner-city communities. While limited in scope, Bullard's research has helped shape the policy debate surrounding environmental justice (Lambert et al. 1996).

Two more research studies played a pivotal role in determining future policies. A widely discussed study examining community demographics near commercial waste treatment, storage, and disposal facilities was conducted by the U.S. General Accounting Office in 1983 for determining the correlation between the location of hazardous waste landfills and the racial and economic status of surrounding communities. Another, often cited empirical study was published in 1987 by the Commission for Racial Justice (CRJ) of the United Church of Christ. The CRJ study had two important components: an analytical survey of commercial waste facilities and a descriptive analysis of uncontrolled toxic waste sites. Both statistical studies were designed to determine the extent to which blacks and other minority groups are exposed to hazardous wastes in their communities.

In August 1994, CRJ released an update of its 1987 study. This update claimed to uncover "even greater racial disparities in the placement of toxic waste sites, despite increased attention to the issue as evidenced by an executive order from President Clinton" (Lambert et al. 1996, p. 3).

Another study deserving mention was published by the National Law Journal in September 1992. Unlike the research discussed above, which focused on the location of industrial and waste facilities, the NLJ study examined racial disparities in Environmental Protection Agency enforcement and remediation procedures.

The crux of environmental justice concerns is that particular communities (chiefly those with racial and ethnic minorities and the poor) have been forced to bear disproportionately the external cost of industrial processes. However, more research is required to clearly demonstrate the validity of such claims.

References

Bullard, R.D. *Dumping in Dixie: Race, Class, and Environmental Quality*. Boulder, Colo.: Westview, 1990.

Center for Policy Alternatives. *Toxic Wastes and Race Revisited*. 1994.

Lambert, T.; Boerner, C.; and Clegg, R. "A Critique of 'Environmental Justice'." *National League Center for Public Interest White Paper*. Vol. 8, no. 1 (January 1996).

Resources from ERIC

Leith, B. "The Social Cost of Sustainability," *Alternatives* 21 (January-February 1995): 18-24. EJ498283.

This article examines the distributive effects of environmental policies designed to promote sustainability. It argues that the obligation to preserve environmental resources for future generations should not undermine the obligation to meet the needs of the current generation if environmental initiatives are to succeed.

Bullard, R.D. "A New 'Chicken-or-Egg' Debate: Which Came First — The Neighborhood, or the Toxic Dump?" *Workbook* 19 (Summer 1994): 60-62. EJ491903.

This article examines racial inequity in the siting of waste treatment, storage, and disposal facilities. It also discusses flaws in several studies that attempt to explain waste facility siting disparities.

Menard, V. "Green Injustice: Who's Winning the Race for Environmental Dollars?" *Hispanic* 7 (November 1994): 18, 20, 22, 24, 26. EJ495389.

Grassroots environmental justice organizations charge that, when funding for projects targeting minority communities is won by multimillion-dollar environmental organizations, community-based environmental justice groups are underfunded and the concentration of toxic waste facilities in minority communities continues to grow. This article also lists the top 10 mainstream environmental organizations and 10 EPA regional offices offering small community grants.

Resources from ENC

Petrikin. J.P., ed. *Environmental Justice*. San Diego: Greenhaven Press, 1995. ENC-014255.

This book, a part of the At Issues series, examines the issue of environmental racism, the disproportionate number of the nation's industrial and waste facilities found in or near minority residential areas.

Ballin, A.; Benson, J.; and Burt, L. *Trash Conflict: A Science and Social Studies Curriculum on the Ethics of Disposal, an Interdisciplinary Curriculum*. Cambridge, Mass.: Educators for Social Responsibility, 1993. ENC-014278.

This guide is designed for teachers in grades six to eight to promote students' understanding of the effects of waste production and disposal and to increase students' feeling of empowerment for the changes they can bring about.

Maser, C. *Resolving Environmental Conflict: Towards Sustainable Community Development*. Delray Beach, Fla.: St. Lucie Press, 1996. ENC-012112.

This book advocates using conflict resolution as a way to deal with the challenges associated with protecting our natural resources. Also provides insights to the necessary philosophical underpinning for practicing a conflict resolution process, called transformative facilitation, to resolve environmental conflicts.

Environmental Justice and Environmental Education

by Bobbi Zbleski, Tom Barrett, Anne Lukosus, Dan Sivek, Sabiha S. Daudi, and Joe E. Heimlich

Traditional environmental education has focused on ecology, environmental issue investigation, and the citizen-action skills needed to understand and influence issue outcomes. Emerging awareness of human health and social problems in racial minority and low-income communities resulting from environmental degradation is expanding the role of environmental education. In order for environmental education to effectively address all populations, curricula need to incorporate environmental justice issues.

What Is Environmental Justice?

Environmental justice is defined by Ruth Wilson, environmental education department editor for *Early Childhood Education Journal*, as the fair and equitable treatment of people of all races, ethnicities, and incomes in regard to development, implementation, and enforcement of environmental laws, regulations, programs, and policies. Fair and equitable practices refer to the idea that no racial, ethnic, or socioeconomic group should disproportionately bear the negative consequences of industrial, municipal, and commercial enterprises. However, statistics indicate that as many as two in five Americans, predominately people of color and low-income families, live where the air is too dangerous to breathe. Other types of ecological destruction can include toxic chemical and solid waste disposal sites, congestion, and excessive noise. Low-income families that live in environmentally degraded communities do not always do so by choice, but often because they lack the means to move to a safer area. Though all people are affected by environmental hazards, children are at a greater risk because a child's body is in a period of rapid growth. Permanent effects from long-term exposure to health hazards are more likely, including brain damage, mental retardation, and respiratory illnesses.

The Environmental Justice Movement

The Environmental Justice Movement is a national, grassroots effort initiated by people of color who are concerned with environmental quality and health hazard issues in their neighborhoods. The movement is a convergence of civil rights and environmental issues and gained national recognition in 1991 during the First National People of Color Environmental Leadership Summit in Washington, D.C. Three types of environmental justice that the movement is focusing on are procedural equity, geographic equity, and social equity. Procedural equity refers to fair scientific and governmental practices. Nonscientific and nongovernmental practices include exclusionary practices, use of only English to conduct meetings in a predominately non-English-speaking community, and public hearings held at inconvenient times and in remote areas. Geographic equity refers to the spatial configuration and proximity of communities to environmental hazards, such as landfills, refineries, lead smelters, and incinerators. Social equity refers to the role of sociological factors, such as race, lifestyle, and class, in environmental decision-making. All three types of inequities overwhelmingly occur in racial minority and low-income communities.

Educating for Environmental Justice

Environmental education can include environmental justice issues by implementing multicultural and urban environmental issue curricula. In the past, environmental education has had a tendency to focus on rural or natural environments and has been taught from a white, middle-class perspective. Though ecological foundations and rural settings are a vital component of environmental education, the need to focus curricula around diverse cultures and the environmental issues pertinent to those groups of people is essential. This is done to ensure that all individuals, regardless of race, ethnicity, or income, minimize their effect on the environment and have the political skills necessary to demand a healthy, intact environment to which all United States citizens are entitled.

There are several suggestions for implementing environmental justice issues and multicultural education into curricula. Karen Warren, outdoors program coordinator and recreational athletics instructor at Hampshire College, suggests using experiential techniques, which can include "giving students profound personal experiences, cooperative learning spaces, multiple learning styles, mindfulness, and vision." Running Grass (1994), executive director of Three Circles Center in Sausalito, California, claims that access to environmental education should be considered an environmental justice issue in and of itself. Encompassing environmental justice issues can be accomplished through multicultural environmental education. Running Grass suggests 10 principles for teaching multicultural education in the classroom. A few of these principles are that multicultural environmental education should: acknowledge that children have different needs based on and shaped by where and how they live; illuminate the idea that all cultures have a relationship with the natural world from which they and other cultures can draw for understanding and inspiration; help to foster in students an awareness, understanding, and acceptance of other cultures and their environmental traditions; promote a society at peace with the natural world and itself; involve family and community institutions directly in the development and implementation of environmental education programs and curricula; and recognize that the health of ecosystems, communities, and individuals are inextricably linked.

While the inclusion of multicultural environmental education is vital in creating an awareness and understanding of environmental justice, Karen Warren cautions that students need to be prepared to expect uncomfortable or dissonant feelings, including guilt, anger, denial, and despair. Environmental justice activities do not offer the same positive experiences or feelings that other forms of environmental education, such as wilderness excursions or trips to nature centers, can generate. There are several steps that can help to prepare students for environmental justice experiences. Despair and empowerment work helps students to turn pain into personal power. The educator creates a safe haven that allows students to express and work through their emotions. Multidimensional processing is used to link environmental justice issues to other disciplines. This helps to ensure that the issue is not forgotten when the class moves to other content areas. Students also gain an awareness of how environmental justice issues are linked to their daily lives and how their decisions can help to alleviate the problem. Outlets for activism are vital to let students know they have opportunities to act on what they have learned.

Resources from ERIC

Bullard, Robert D. "Overcoming Racism in Environmental Decision-Making." *Environment* 36 (May 1994): 10-20, 39-44. EJ487140.
    This article is a historical overview of the environmental justice movement and recommends five principles of environmental justice to promote procedural, geographic, and social equity. This resource also includes Executive Order No. 12898, concerning federal actions to address environmental justice issues.

Grass, Running. "Towards a Multi Cultural Environmental Education." *Multi Cultural Education* 2 (Fall 1994): 4-6. EJ494140.
    This article presents a brief rationale for multicultural environmental education, provides an outlined multicultural critique of environmental education, and offers some draft principles of a multicultural environmental education.

Potter, John F. "Environmental Justice: Co-Evolution of Environmental Concerns and Social Justice." *The Environmentalist* 14 (Summer 1994): 121-29. EJ496811.

This article is a historical perspective that describes the co-evolution of environmental concerns with civil rights that has found overlap in the environmental justice movement. It also discusses the implications for decision making and protecting both environmental quality and civil rights.

Infusing Environmental Education: Some Models

Environmental education can help an educator teach across the curriculum. One of the dominant means by which this is done is thematically. By using environmental themes, students explore and develop questions that they must answer in order to proceed. The means by which they get the answers is through the application of knowledge and skills from the disciplines. For example, a scavenger hunt for different plants not only can expose students to the plants themselves, but also can provide lessons on cartography or geography. Or a study of a pond can provide not just a study of the science of the pond, but also a study of the community of the pond, the human use of the pond, mathematics (slope, angles, algebra), mapping, cultural studies, and civics or social studies (zoning). Environmental themes can encourage students to use diverse cognitive skills that provide a more well-rounded environmental education experience.

Locally, there are many providers of environmental education, such as parks, zoos, nature centers, museums, science centers, waste management programs, utility companies, and others who offer resources for teachers to use in incorporating environmental education into the classroom. Many of these programs correlate their activities with state or national standards or proficiencies. There are also national programs that provide well-tested materials for teachers to use in their classrooms. Water monitoring, weather monitoring, and other environmental monitoring data are often products of teacher-led, classroom-based, environmental education projects that are coordinated on a regional, national, or international scale.

This section of the collection points educators toward resources and programs that provide a means of infusing environmental education into the curriculum or creating a theme by which the curriculum can be taught.

Exploring Environmental Education in the Schoolyard Habitat

by Mark A. Miller, Sabiha S. Daudi, and Joe E. Heimlich

One of the renewed trends in environmental education is the belief that subjects and issues intrinsic to the study of our natural world can be explored within the surroundings in which a school is located. During the past two to three decades, this belief has crystallized into the schoolyard habitat movement, a contemporary trend with roots in the historic "outdoor education" movement of the 1950s. In an article for *Early Childhood Education Journal*, Mary Rivkin outlined the schoolyard habitat movement and its importance to environmental education (1997). She noted that many older Americans still remember outdoor play as a treasured part of their early experiences, but most of today's youth have lost the natural habitat as a place to learn. Urbanization, industrialization, predominance of automobiles, deteriorated social conditions, and the side effects of technology have all contributed to the isolation of contemporary youth from the natural environment.

Benefits of Learning in a Schoolyard Setting

The reasons for exploring natural habitats with children are many:

- According to the "biophilia hypothesis," humans have an inherent need for affiliation with natural environments, just as they have an inherent need for contact with other humans.
- Children are multi-sensory, physical beings and benefit physically, cognitively, and emotionally from an interplay with wind and water, sights and sounds, plants and animals, running and shouting.
- Research has shown that children who do not play in natural habitats are unaware of and possibly do not care about the plants and animals that live around them.
- Children have more positive social relationships and more creative play in natural environments.
- Nearly every major subject matter can be taught in a schoolyard setting, and much of the learning takes on an experiential aspect.

The Schoolyard Habitat Movement

A number of schools have had grounds improvement projects since the resurgence of environmental activism in the 1970s. National environmental education programs, such as Project Learning Tree and Project WILD, have fostered these improvements. Learning Through Landscapes, a highly successful national program in Great Britain, has inspired similar programs in Canada and Sweden and has provided fresh impetus to the schoolyard improvement efforts in the United States (Rivkin 1997).

Using the schoolyard for learning provides a laboratory for exploring all disciplinary-based courses, as well as multi- and trans-disciplinary based curricula. The major difference between the schoolyard habitat movement and the older idea of outdoor education resides in the nondisciplinary view. Taking students outdoors, whether into a land laboratory or onto the sidewalk, can enhance learning, not just learning about natural science.

Many states and school districts in the United States have recognized the validity of developing and adopting curricula for teaching in a schoolyard setting. Many schools establish or build on existing surroundings that allow students to participate in eco-system exploration and management. Organizations have been formed that are devoted solely to school grounds enhancement or related sponsoring programs, while more traditional wildlife con-servation groups now are viewing schoolyards as places to directly inform children about their natural heritage and encourage them in its preservation.

Reference

Rivkin, M. "The Schoolyard Habitat Movement: What It Is and Why Children Need It." *Early Childhood Education Journal* 25, no 1 (1997).

Resources from ERIC

Dunbar, Terry. "Acting Locally: On Site Science." *Green Teacher* (June-September 1994): 18-19. EJ516942.

Describes an innovative science program that focuses on elemen-tary and secondary students and trains teachers to use their own schoolyards as environmental laboratories.

Gosselin, Heather, and Johnson, Bob. "Amphibian Oasis: Designing and Building a Schoolyard Pond." *Green Teacher* (June-September 1996): 9-12. EJ540032.

Building a pond in a school yard is a rewarding way to help boost local populations of amphibians, to increase the natural value of school grounds, and to serve as a locale for observing the life cycles of plants, invertebrates, and amphibians. This article outlines impor-tant considerations in designing and building a pond from siting through maintenance.

Krupa, Karen. "The Abundance of Nature's Imagination: Schoolyard Naturalization as an Inspiration for the Arts." *Green Teacher* (April-May 1994): 16-17. EJ496879.

An artist and parent describes the potential for a schoolyard re-naturalization to be integrated into art curricula and some art activi-ties that have been inspired by planting in a schoolyard.

"Schoolyard Habitats: Learning Locally. Facilitator Training." Pro-ceedings of a workshop presented at the annual conference of the

North American Association for Environmental Education, Burlingame, Calif., 1-2 November 1996. ED403127.

This workshop presented components of successful programming, including forming a broad-based team, site inventory and mapping techniques, managing the site for diversity, project and program evaluation, and tips on working with schools. It also provides examples of real school projects and ideas for involving the community and other partners.

Resources from ENC

Evergreen Foundation. http://www.evergreen.ca/home.html. ENC-002114.

This website provides users with information about the foundation's efforts to connect people with nature by developing and enhancing healthy, natural environments in schools and communities across Canada.

Hogan, Kathleen. *Eco-Inquiry: A Guide to Ecological Learning Experiences for the Upper Elementary/Middle Grades.* Dubuque, Iowa: Kendall/Hunt, 1994. ENC-008947.

This ecology curriculum guide is intended to build students' understanding of three fundamental ecological processes (food webs, nutrient cycling, and decomposition) in the students' local environment and to help them understand the positive and negative effects of their own individual actions on ecosystems.

Slattery, B.E., et al. *WOW! The Wonders of Wetlands.* Bozeman, Mont.: The Watercourse, and St. Michaels, Md.: Environmental Concerns, 1995. ENC-004983.

This activity book, designed for grades K-12, provides hands-on learning experiences on wetlands. Some activities deal with establishing a wetland in a schoolyard environment and focus on the interaction between wetlands and humans, including social and political issues.

Ballbach, J., ed. *Ohio Sampler: Outdoor and Environmental Education.* Newark: Environmental Education Council of Ohio, 1995. ENC-005882.

This book contains environmental and outdoor education activities used by practitioners. The activities are divided into classroom activities, schoolyard activities, and outdoor nature activities.

Earth Notes: For Educators, Grades K to 6. Washington, D.C.: U.S. Environmental Protection Agency, 1992. ENC-003824.

This journal provides an open forum for the exchange of ideas, comments, and brief essays. It also features an article on the Urban Ecology Education Project, a program that solicits community support of schoolyard ecology projects.

School Garden-Based Environmental Education: A Case Study

by Mark A. Miller, Joe E. Heimlich, and Sabiha S. Daudi

Environmental education practitioners have strongly proposed a cross-disciplinary approach for bringing environmental education into the classroom. The nature of teaching and learning materials, those currently established and those under review, enables educators to achieve this. However, one reservation often expressed by educators is the lack of time and resources. The following case study suggests some answers through garden-based activities that use the school garden as a learning context and that highlight environmental concerns and issues within the school boundaries.

Waynewood Elementary School, Virginia

Mrs. Bowman's second-grade class at Waynewood Elementary in Alexandria, Virginia, was an ethnically diverse group of 19 students representing eight different countries. As one of the Fairfax County schools that offered English as a second language to recent immigrants, children from around the world were placed in classes with children born and raised in the United States. Through this total immersion approach to learning English, the new children generally "graduated" within three years at an excellent fully functional level. In this type of classroom, Mrs. Bowman found a number of challenges above and beyond basic communication. Foremost, there was the question of how to teach a myriad of subjects in a way that all the students could relate to and understand despite their divergent cultures and backgrounds.

The most substantial and engaging way in which Mrs. Bowman succeeded in reaching her second-graders was by establishing a school garden. With the help of parent and community volunteers and other school staff, they created a school garden project modeled on the "garden in every school" initiative begun by Delaine Eastin, State Superintendent of Public Instruction in California. It was Ms. Eastin's desire to create opportunities for children to make healthier food choices, to participate in experiential learning, and to develop a deeper appreciation for the environment, the community, and each other through school gardens.

A study by the California Children's Five a Day — Power Play Campaign suggested actively involving children in gardening to increase their fruit and vegetable consumption. At the same time, teachers could use garden-based education effectively to build bridges between school and community, promote intergenerational knowledge transfer, develop environmental awareness in students by caring for a living environment, provide opportunities for cultural exchange, and build student life skills.

Every subject that Mrs. Bowman needed to cover in her curriculum was integrated into the school garden project at Waynewood. Understanding the nature of soil, weather, interactions with other organisms, and the necessary ingredients for successfully growing plants was a critical part of establishing the garden. Students were introduced to the scientific methods of investigation by becoming vegetable detectives: researching the world origin of their vegetable, observing the life cycles of their plants, analyzing which nutrients were found in their vegetable and how those nutrients are used in human bodies, and describing the parts of their plant that are edible. Mathematics was essential for the design of the garden and subsequent measurement of plant growth. The history and cultural significance of each garden plant were considered. Stories about gardens and the world outside the classroom were regularly read by teacher and students. Art class consisted of drawing the plants and other living creatures found in the garden. The end of the school year culminated in a festival held in the school cafeteria. All of Mrs. Bowman's students not only were able to proudly share the garden bounty with the other students and staff of the school; they also were able to communicate the particulars of each fruit or vegetable being eaten and the environmental factors associated with them.

Outcomes

This case study presents a methodology where the teacher successfully provided a learning environment to students who not only were able to experience real-life issues but also gained insights by being involved. She taught environmental education using a cross-disciplinary approach and simultaneously taught such core subjects as math, science, and geography.

Resources from ERIC

Demas, S. "School Gardens and Environmental Education." *Nature Study* 32, no, 3 (1979): 3-5. EJ200262.

This article discusses the use of garden projects to teach environmental concepts. Also described are several youth garden programs and home/school programs in existence throughout the country.

Monk, S.K. "Integrated Curriculum in a Tiny Texas Garden." *Dimensions of Early Childhood* 23 (Summer 1995): 8-9. EJ510569.

This paper describes how a first-grade class in East Texas learned to garden in a project that encourages children to plant seeds and watch their plants mature while developing skills in mathematics, science, reading and writing, and interpreting a calendar.

Greenhalgh, L. "An Infant School Project." *Environmental Education* 50 (Autumn 1995): 11. EJ546447.

This article describes a wildlife garden developed on school grounds that converted a large, derelict quadrangle into an outdoor classroom used by the staff across the curriculum. The area was divided into various habitats, including ponds, mini-woodlands, and an herb garden.

Hanscom, J.T., and Leipzig, F. "The Panther Patch: A Far North K to 6 Gardening Project." *Green Teacher* (April-May 1994): 10-13. EJ496877.

This article outlines the development of an urban elementary school garden where children learn science and responsibility for the environment.

Moore, R.C. "Children Gardening: First Steps Towards a Sustainable Future." *Children's Environments* 12 (June 1995): 222-32. EJ509046.

Children's gardening is introduced within the broader frame of sustainable development, regenerative design, and biodesign. Gardening in the primary grades is proposed as one of the most feasible pedagogical approaches for ensuring a daily learning experience that provides contact with nature.

Resources from ENC

Life Lab Science Program. *Life Lab Garden Log*. Santa Cruz, Calif.: Let's Get Growing! 1993. ENC-003373.

Life Lab is an interdisciplinary program of life, Earth, and physical science in which students learn science concepts by building tools, testing ideas, and watching changes in the world around them. A class garden and hands-on activities form the core of the program, encouraging students to cooperatively investigate life cycles, weather, animals, habitats, and more.

National Gardening Association. *GrowLab: Activities for Growing Mind*. Burlington, Vt., 1990. ENC-001518.

This program, developed for grades K-8, provides a context in which students can use their own questions and ideas to inspire hands-on investigations of the natural world, explore key plant science and environmental concepts, and gain confidence in their ability to do science.

Jurenka, N.A., and Blass, R.J. *Cultivating a Child's Imagination Through Gardening*. Englewood, Colo.: Teachers Ideas Press, 1996. ENC-008702.

This book, written for grades K-6, is organized around a theme of gardening and uses children's literature to introduce students to multicultural literature, ecology, and the impact of plants on the environment, world economics, and politics.

Raymond, D. *Down to Earth Gardening Know-How for the 90's: Vegetables and Herbs*. Pownal, Vt.: Storey Communications, 1991. ENC-003367.

This gardening manual incorporates the lifestyle changes of the 1990s into the author's 40 years of gardening experience. Also included is information for planning, fertilizing, cultivating, and harvesting a garden.

Window on the World: Classroom Gardening

by Mark A. Miller, Sabiha S. Daudi, and Joe E. Heimlich

Many forces in our modern society have alienated America's youth from the natural environment. The dramatic increase in children being raised in suburban and urban settings has separated them from opportunities to understand the world and themselves through the constant exploration of nature. Research has shown myriad physical, emotional, and cognitive benefits to children who are able to discover and play in natural environments (Rivkin 1997). The schoolyard habitat movement that has blossomed out of an understanding of these benefits is gaining strength in the United States and abroad; however, a number of schools are unable to participate in creating a habitat at their schools due to significant impediments. A few reasons a school may be unable to establish an area for students to explore outside include:

- Lack of any open land.
- Lack of necessary resources.
- Unfamiliarity of school staff with initiating a garden or habitat.
- Safety concerns.
- Lack of administrative support.

An alternative to the schoolyard habitat or land laboratory is the approach of classroom gardening. Flowers or vegetables —

planted in pots, trays, tubs, or buckets — can create experiences for a host of teachable concepts. Teachers' efforts using class-room gardening vary from flats of seedlings on windowsills to complex growing stations in the classroom. Alternatives available to educators include such activities as "fast plants" and two-liter bottle terrariums.

The process of classroom gardening begins with teacher iden-tification of goals and outcomes for the students through projects that can be completed within the classroom. A number of sources explain curriculum activities, lessons, and experiments for the teacher to use, along with helpful hints on generating enthusiasm within the class; how to garner school and local support; mainte-nance of classroom gardens; uses of the garden plants grown; and such related activities as crafts, composting, and vermiculture. The resources listed in this information sheet give specific exam-ples of the types of indoor gardens that can be established.

Educators from various arenas have developed classroom cur-ricula that open up the outside world for student exploration through the activity of gardening. The simple act of working with soil or planting a seed becomes magical for many students unfa-miliar with gardening. Learners can then dig into the relationship between plants and the many products, foods, and medicines derived from them. Examining the process of gardening within different ethnic groups can bring about a better realization of self, as well as a better appreciation for others. The subject of plants and their cultivation leads students toward an understanding and appreciation of life in a holistic frame. In other words, classroom gardening can open a window on the world.

Classroom gardening also provides teachers with a context for hands-on research into a wide variety of basic subject areas. The activities inherent in gardening become an interdisciplinary vehicle for the teaching of skills in science, mathematics, social studies, language, health, and art. While fulfilling mandated cur-riculum requirements, these activities facilitate socialization and other affective outcomes as well.

Reference

Rivkin, M. "The Schoolyard Habitat Movement: What It Is and Why Children Need It." *Early Childhood Education Journal* 25, no. 1 (1997).

Resources from ERIC

Jurenka, N.A, and Blass, R.J. *Beyond the Bean Seed: Gardening Activities for Grades K-6*. Englewood, Colo.: Teacher Ideas Press, 1996. ED395814.

This book connects gardening with literacy and children's literature and is designed for adults who work with children, including classroom teachers, horticulturists, arboretum and botanical garden education directors, librarians, garden center teachers, and camp counselors.

Larson, J. "Environmental Education Activities K-12: Gardens for Young Growing Lives." *The Green Pages* (January-February 1997): 20-24. EJ541737.

Describes several gardening activities that can be kept simple or used as a foundation for more in-depth projects. Activities include setting up an indoor garden spot; making compost, which helps students understand the terms *decompose* and *compost*; watching plants drink, in which students measure water movement in plants; making herb gardens; and drying herbs.

Pranis, Eve, ed. *Growing Ideas: A Journal of Garden-Based Learning*. vols. 1-4. Burlington, Vt.: National Gardening Association, 1990-1993. ED367544.

Each issue in this series of journals provides instructional ideas, horticultural information, and a forum for exchange among teachers using classroom gardening to stimulate learning. Ideas in each issue are separated into three sections: "Green tips," "Exchange," and "Resources."

Pranis, E., and Hale, J. *GrowLab. A Complete Guide to Gardening in the Classroom*. Burlington, Vt.: National Gardening Association, 1991. ED366512.

Having a garden in the classroom offers students a chance to explore plant life cycles and provides a hands-on context for teaching a wide variety of basic skills in science, mathematics, social studies, language arts, health, and fine arts. This teacher guide is designed to help teachers in grades K-8 establish and maintain a garden in their classroom.

Resources from ENC

Jurenka, N.A., and Blass, R.J. *Cultivating a Child's Imagination Through Gardening*. Englewood, Colo.: Teacher Ideas Press, 1996. ENC-008702.

This book, written for grades K-6, is organized around a theme of gardening and uses children's literature to introduce students to multicultural literature, ecology, and the effect of plants on the environment, world economics, and politics.

Petrash, C., and Cook, D. *Earthways: Simple Environmental Activities for Young Children*. Beltsville, Md.: Gryphon House, 1992. ENC-006274.

This activity book focuses on enhancing children's environmental awareness. The book is organized according to the four seasons. Each section contains suggestions to make the classroom more environmentally friendly (recycling, composting, reusing materials, and others).

Rigby. *The Wonderful World of Plants*. Kit. Science Alive Series. Auckland, N.Z.: Shortland Publications, 1993. ENC-008289.

This kit introduces students to the characteristics and life cycles of plants. The aim of the Science Alive series is to provide a comprehensive elementary curriculum that develops children's willingness and ability to investigate, understand, and explain scientific phenomena as a means of making better sense of their world.

Woolfitt, G. *Sow and Grow*. New York: Thomson Learning, 1994. ENC-005112.

This hands-on activity guide, designed for grades 2-4, uses full-color photographs and engaging activities to teach the basics of plant life and to reinforce scientific principles. In sample activities, students study plant parts, observe how a seed germinates, and learn how to grow plants from seeds, bulbs, tubers, and cuttings.

The Importance of Energy Education in Environmental Education

by Jennifer Folsom, Bonnie Koop, Dan Sivek, and Sabiha Daudi

Energy is the ability to do work or to change things. We use electricity to power our homes; we use gasoline and oil that come from fossil fuels to drive our cars, and we depend on energy from food for our bodies to function properly. "Energy is the underlying currency that governs everything humans do with each other and with the natural environment that supports them" (KEEP Conceptual Guide 1996). However, because energy is so prevalent in our lives, it is easy to take it for granted. Energy education is essential for a growing population faced with finite resources and serious environmental issues.

Goals of Energy Education

Energy education programs are designed to increase understanding of how we and other things on Earth use energy, to analyze issues that arise out of energy resource development and use, and to explore resource management practices that address these issues. Energy education strives to teach people to apply critical-thinking skills in evaluating energy issues and to become a knowledgeable, active citizenry that uses energy responsibly.

Making the Connection

Making the connection between energy issues and environmental issues is a logical and necessary one. "While the 'Energy Crisis' has faded from the headlines . . . environmental and economic issues [related to energy use] have not" (Snyder et al. 1994). Incorporating energy education into the context of environmental education complements the goals of environmental education. Through energy education, students learn basic ecological concepts and gain citizen action skills relevant to environmental issues. Almost any environmental issue can be connected to energy use, (for example, global warming, acid rain, and mining issues). "Environmental education . . . should constitute a comprehensive lifelong education, one responsive to changes in a rapidly changing world." (Tbilisi Declaration, 1977). The North American Association for Environmental Education recognizes the importance of energy education to environmental literacy. *Excellence in Environmental Education: Guidelines for Learning (K-12)* consistently incorporates energy concepts into themes to ensure comprehensive coverage of energy within environmental education lessons. To strengthen people's understanding of environmental issues, it is essential that they have a good understanding of the importance of energy and the pivotal role it plays in those issues.

Energy Education Infusion

Every subject and grade level has the opportunity to teach about the relevancy and importance of energy. For example, science teachers can emphasize energy flow when teaching about food chains. Through energy education, students can explore how energy has shaped past and current environmental, political, and sociological issues. Students can make history come alive by interviewing grandparents or nursing home patients and learning firsthand how past energy use compares to the present. Home energy use audits will be useful to Family and Consumer Education students when they learn ways to save money in their household budgets by conserving energy. Infusing energy education into an existing curriculum also can be used as a tool to meet many objectives, such as critical-thinking skills. By understanding how we use energy and its costs and benefits, students will gain important thinking and decision-making skills. Energy education is integral to improving our students' environmental literacy.

Energy Education Sources

The following resources can be located on the databases ERIC and ENC. They are excellent for locating materials to begin the incorporation of energy education into existing curricula. Some resources provide evaluations of existing programs in an effort to help determine which sources and guides are needed and appropriate. Others are actual curriculum and activity guides. Access to these databases is available at local libraries, universities, and online. Internet access to these resources is available through the databases of Educational Resources Information Center (ERIC) and Eisenhower National Clearinghouse (ENC) collections.

Resources from ERIC

Glass, Lynn W., ed. *Energy Education: Focus on Excellence.* vol. 3. Washington, D.C.: National Science Teachers Association, 1985. ED328404.

This resource is useful for new program modeling. It includes descriptions of eight successful energy education programs and the reasons for each curriculum's selection. The goals and criteria for excellence in science education programs also are provided.

Schlene, Vickie J. "Teaching About Energy." *Social Education* 56 (January 1992): 8. EJ447828.

This article contains a list of many of the documents available on ERIC that are useful for energy education, as well as guidelines and suggestions for infusing energy education in the curriculum. Examples of successful state and federal programs and other outstanding curriculum materials are included.

Blueprint for Success: An Energy Education Unit Management Plan. Reston, Va.: National Energy Education Development Project, 1995. ED390681.

This is an interdisciplinary resource emphasizing knowledge building and skills enhancement. Teamwork and peer teaching are accented. Descriptions, time lengths, intended grade levels, ability level adjustments, and teacher/student evaluations for 41 activities are included.

Resources from ENC

Altman, Paula, comp. *Energy Education Resources: Kindergarten Through 12th Grade*. Washington, D.C.: National Energy Information Center, 1995. ENC-003273.

This resource guide contains a list of free and inexpensive materials available to teachers and students. Extensive contact information for each source is provided, as well as a comprehensive and easy-to-use index. Direct Internet access is available at: http://www.eia.doe.gov/bookshelf.html

Snyder, Will, and the 1994 National 4-H Energy Education Review Team. *Educating Young People About Energy for Environmental Stewardship: A Guide to Resources for Curriculum Development with an Emphasis on Youth-Led, Community-Based Learning*. Chevy Chase, Md.: National 4-H Council, 1994. ENC-010475.

This K-12 resource guide is designed to help develop energy/environmental stewardship in students. Guidelines for infusing energy/environmental stewardship education are discussed, including techniques, resource availability, and grade-level guidelines. The included activities and curricula were evaluated on content, ease of use, and educational techniques. The contacts for obtaining resources are listed.

EnergyNet Community Web. ENC-012957. http://www.energynet.net

This website is designed to facilitate energy education by allowing students and their teachers in grades 6-12 to use telecommunications. Teamwork, community and school involvement with energy conservation and education, peer networking, and expert contacts are the main emphases. Students are able to post the results of their work on the site, link to other appropriate sites, and access Energy-Net support staff and energy experts. Development conferences are available for teachers.

Earth Systems Education: Learning About Earth as a System

by Sabiha S. Daudi and Joe E. Heimlich

The Earth as seen from space appears to be a complete unit independent of any links with other planets of the solar system, the galaxy, or the universe. This apparently self-reliant planet, when explored closely, is made up of systems that are interrelated and interconnected to keep many subsystems running — systems that collectively form the basis of life on Earth. Learning about the art and science of Earth systems is one of the intellectual challenges for the future and forms the necessary foundation for sustainable management of our planet.

During the past several decades, under the increasing pressure of public concerns about air and water pollution, nuclear waste disposal, the ozone hole, climate change, and other issues, the scientific community has realized how interrelated are the components that once were viewed as separate parts of the Earth system. While the critical parts — processes in the biosphere, hydrosphere, atmosphere, and lithosphere — are studied in detail within the boundaries of traditional disciplines, how the parts combine and interact is the key to understanding how our home planet works, its past history, and its likely future. The idea of *Earth System Science* has emerged as a new and revolutionary unification of the science of our planet.

Earth Systems Education (ESE) plays an important role in integration of science disciplines in numerous academic institutions and is represented by programs on all continents. Earth Systems Education can be approached formally on the K-12 level as well, providing a practical approach to the modernization of science curricula. It also contributes to global science literacy by helping students gain a richer understanding of the nature of the planet on which they live. As students learn to make informed choices in their science investigations, they learn skills for approaching real-world issues as well.

Mayer (1993) has identified four goals for Earth Systems Education that focus on developing:

- *scientific thought*: to be able to understand the nature of scientific inquiry using the historical, descriptive, and experimental processes of Earth sciences;
- *knowledge*: to describe and explain Earth processes and features and anticipate changes in them;
- *stewardship*: to respond to environmental and resources issues in an informed way; and
- *appreciation*: to develop an aesthetic appreciation of the Earth.

These goals preceded, and now support, those of the National Science Education Standards. Like the Standards, ESE encourages:

- a hands-on, minds-on investigative approach;
- collaborative, not competitive, learning;
- integration of science disciplines through the investigation of important science questions;
- local learning with global applications;
- capturing the excitement and fun of learning about planet Earth.

Earth Systems Education is a grassroots movement for positive curriculum restructure in science. It begins with individual teachers in their classrooms, implementing small changes as they learn how to adapt their current methods and materials to a systems approach. A guidebook of resources to assist such efforts is *Science Is a Study of Earth* (Mayer and Fortner 1995).

ESE is guided by a simple framework of seven understandings developed by science teachers, teacher educators, and scientists to help teachers assess their own progress in implementing a science curriculum with all the essential elements. A curriculum organized to fulfill these understandings will be a relevant, exciting means of building global science literacy.

Resources from ERIC

Activities for the Changing Earth System. Earth Systems Education. Middle and High School Activities. Columbus, Ohio: ERIC/CSMEE, 1993. ED382444.

This book is intended to help teachers fill the need for children and future leaders to understand issues of global change and the science that lies behind them. The book focuses on important changes that are occurring in several of Earth's subsystems as a result of natural and human-caused factors.

Mayer, V.J., and Fortner, R.E., eds. *Science Is a Study of Earth: A Resource Guide for Science Curriculum Restructure.* Greeley: University of Northern Colorado, and Columbus: Ohio State University School of Natural Resources, 1995. ED391638.

The 250-page loose-leaf guide provides the rationale for ESE; examples of how it fits with the scope and sequence of science in elementary, middle, and high schools; worksheets for examining and modifying existing instructional activities; suggestions for evaluation; list of resources; and sample activities for all grade levels.

Fortner, R.W., and Boyd, S. "Infusing Earth Systems Concepts Throughout the Curriculum." Paper presented at the annual meeting of the National Association for Research in Science and Technology, San Francisco, 1995. ED386391.

The Program for Leadership in Earth Systems Education (PLESE), a teacher enhancement program sponsored by the National Science Foundation in 1990-94, was a coordinated effort to infuse Earth Systems concepts throughout the K-12 science curriculum across the United States. Characteristics of the program are reviewed in this paper and the evaluation of its components are discussed.

Mayer, V.J. "Earth-Systems Science: A Planetary Perspective." *Science Teacher* 58 (January 1991): 35-39. EJ419096.

Presented here are the reasons why the Earth sciences must play a major role in curriculum renovation. Philosophical, methodological, and conceptual contributions to K-12 curriculum content are discussed.

Resources from ENC

Brown, S.; Crosby, D.; and Jax, D. *Earth System Science Activities.* Columbus, Ohio: Bexley Middle School, 1995. ENC-005018.

   The activities in this book, designed for grade 6-12, were developed using the Earth Systems Education Framework. This framework was designed to be used by teachers to select and develop units of study that focus on the appreciation, stewardship, and understanding of the Earth and its resources through activities that engage students in hands-on learning experiences.

Mayer, V.J. *Earth Systems Education.* Columbus, Ohio: ERIC Clearinghouse for Science, Mathematics and Environmental Education, 1993. ENC-012263.

   This article, written for K-12 educators and produced as a digest produced by ERIC/CSMEE, describes the Earth System Education Program at Ohio State University and the University of Northern Colorado. Topics include technological advances that have increased our understanding of planet Earth; the PLESE (Program for Leadership in Earth Systems Education) teacher training program; and integrating Earth systems education into the curriculum.

Miller, H., and Sheaffer, A., eds. *Great Lakes Instructional Materials for the Changing Earth System (GLIMCES): An Earth Systems Education Effort of the Ohio Sea Grant College Program and the Ohio State University.* Columbus, Ohio: Ohio Sea Grant Publications, 1995. ENC-004547.

   Designed for grades 5-12, this book presents activities that help students to understand how the Great Lakes Region will be affected by global climate change. The material is based on scientific predictions for changes in water resources, ecological relationships, geology, critical environments, and commercial and recreational use of the lakes.

Harden, B.J. *Planet Earth Home Page.* ENC-008729. www.tidusa.com/Planet_Earth/index.htm

Planet Earth Home Page (PEHP) is a virtual library of resources available on the Web. It currently contains 13 rooms with 17 shelves of links in each room. The education room contains links to professional development opportunities, graduate schools, and other programs, as well as educational sites for educators at all levels.

Acknowledgement: thanks to Rosanne Fortner for providing valuable insights and comments on the contents of this information sheet.

Earth Systems Education: Underlying Understanding

by Sabiha S. Daudi and Joe E. Heimlich

There have been tremendous advances in the understanding of planet Earth in the recent decades, in part through the use of satellites in data gathering and supercomputers for data processing. As a result, Earth scientists are reinterpreting the relationships among the various science subdisciplines and their mode of inquiry.

While Project 2061 was finalizing its content goals and the National Science Education Standards were still in the planning stages, some scientists and educators began to voice their concerns about the position of the sciences of planet Earth in the nation's K-12 curriculum. Their intent was to ensure that in curriculum restructuring efforts, more student experiences and content regarding modern understandings of our home planet would be provided. A basic premise was that any physical, chemical, or biological process that citizens must understand to be scientifically literate can be taught with greater relevance in the context of the Earth subsystems. A further premise was that "learning locally," using here-and-now examples, can assist students in "applying globally" when environmental decisions are made that affect the planet.

A group of leaders in science education (teachers and teacher educators) and the geosciences met first in 1988 to consider what every graduating high school senior should know about Earth. An initial framework of ten "Understandings" (with greater depth than facts and concepts) was developed. Through subsequent discussions with teachers and Earth science educators at meetings of the National Science Teachers Association (NSTA), and a special focus group of scientists and educators convened under NSF's Program for Leadership in Earth Systems Education (1990-95), the final Framework for Earth Systems Education emerged.

It was the geoscientists (geologists, meteorologists, oceanographers, and astronomers) on the development team who insisted that these elements alone did not account for why they chose science as a career. Thus they added Understanding #1, to be first on any list, relating the wonder of Earth and the value of it. Then, because of the beauty and value of Earth, they felt that stewardship of the planet belonged in the next priority position. It is these two understandings, and the succinctness of the total list, that make the Framework for Earth Systems Education unique and appropriate guidance for all of science education, as well as confirming the goals of environmental education.

Framework for Earth Systems Education

Understanding #1: Earth is unique, a planet of rare beauty and great value.

- The beauty and value of Earth are expressed by and for people through literature and the arts.
- Humans' appreciation of planet Earth is enhanced by a better understanding of its subsystems.
- Humans manifest their appreciation through their responsible behavior and stewardship of its subsystems.

Understanding #2: Human activities, collective and individual, conscious and inadvertent, affect planet Earth.

- Earth is vulnerable, and its resources are limited and susceptible to overuse or misuse.
- Continued population growth accelerates the depletion of natural resources and destruction of the environment, including other species.
- When considering the use of natural resources, humans first need to rethink their lifestyles, then reduce consumption, then reuse and recycle.
- By-products of industrialization pollute the air, land, and water; and the effects may be global, as well as near the source.

Understanding #3: The development of scientific thinking and technology increases our ability to understand and use Earth and space.

- Direct observation, simple tools, and modern technology are used to create, test, and modify models and theories that represent, explain, and predict changes in the Earth system.
- Historical, descriptive, and empirical studies are important methods of learning about Earth and space.
- Scientific study may lead to technological advances.
- Regardless of sophistication, technology cannot be expected to solve all of our problems.
- The use of technology may have benefits, as well as unintended side effects.

Understanding #4: The Earth system is composed of interacting subsystems of water, rock, ice, air, and life.

- The subsystems are continuously changing through natural processes and cycles.
- Forces, motions, and energy transformations drive the interactions within and between the subsystems.
- The Sun is the major external source of energy that drives most system and subsystem interactions at or near the Earth's surface.

- Each component of the Earth system has characteristic properties, structures, and compositions that may be changed by interactions of subsystems.
- Plate tectonics is a theory that explains how internal forces and energy cause continual changes within Earth and on its surface. Weathering, erosion, and deposition continuously reshape the surface of the Earth.
- The presence of life affects the characteristics of other systems.

Understanding #5: Planet Earth is more than four billion years old, and its subsystems are continually evolving.

- Earth's cycles and natural processes take place over time intervals ranging from fractions of seconds to billions of years.
- Materials making up planet Earth have been recycled many times.
- Fossils provide the evidence that life has evolved interactively with Earth through geologic time.
- Evolution is a theory that explains how life has changed through time.

Understanding #6: Earth is a small subsystem of a solar system within the vast and ancient universe.

- All material in the universe, including living organisms, appears to be composed of the same elements and to behave according to the same physical principles.
- Nine planets, including Earth, revolve around the sun in nearly circular orbits.
- Earth is a small planet, third from the Sun in the only system of planets definitely known to exist.
- The position and motions of Earth with respect to the Sun and Moon determine seasons, climates, and tidal changes.
- The rotation of Earth on its axis determines day and night.

Understanding #7: There are many people with careers that involve study of Earth's origin, processes, and evolution.

- Teachers, scientists, and technicians who study Earth are employed by businesses, industries, government agencies, public and private institutions, and as independent contractors.
- Careers in the sciences that study Earth may include sample and data collection in the field and analyses and experiments in the laboratory.
- Scientists from many cultures throughout the world cooperate and collaborate using oral, written, and electronic means of communication.
- Some scientists and technicians who study Earth use their specialized understanding to locate resources or predict changes in Earth systems.
- Many people pursue avocations related to planet Earth processes and materials.

This framework provides the teachers with the necessary information that they need to plan and develop their curriculum and to coordinate it with the science activities.

Resources for Environmental Education

There are literally tens of thousands of teaching resources for educators on environmental topics, issues, and environmental education in general. Yet, for many educators, finding these resources — or even more important, finding the time to find the resources — is a major challenge. The emergence of the Web has certainly assisted in expediting the process, but effective searching requires knowing both what one is looking for and knowing the various ways of finding it.

There are some resources available through the EETAP Resource Library to help teachers become better at searching for resources. *Finding Resources on the Internet, Evaluating Content of Web Sites,* and *Evaluating the Structure of Web Sites* are all available at no charge in a PDF (downloadable) format. Also on this site is the *Environmental Education Information Providers Directory*, which lists many of the online sites that have resources available, some free, some for sale.

There are also many resources in newsletters, magazines, and publications of all sorts. Teachers often overlook some excellent sources of activities or ideas because many of the resources they might find in a search are more "academic" journals. It is often impossible to tell from a listing of a journal if that journal publishes basic research or if it includes applications, evaluations of programs or projects, or teaching activities.

This section of the collection provides some guidance for educators on some of the many resources available for understanding, incorporating, and practicing good environmental education in the classroom. In these articles, we've taken the opportunity to do some of the searching, sifting, and synthesizing for teachers and tried to make it easier to find the resources teachers want.

What Is ERIC?

Adapted by Liz Barringer-Smith

from "All About ERIC" by Lynn Smarte and the Access ERIC staff

The Educational Resources Information Center (ERIC) is a federally funded, nationwide information network designed to provide individuals with ready access to education literature. The ERIC system is a part of the U.S. Department of Education's National Library of Education and consists of 16 ERIC Clearinghouses, 11 Adjunct Clearinghouses, and additional support components.

The mission of the ERIC system is "to improve American education by increasing and facilitating the use of educational research and information to effect good practice in the activities of learning, teaching, educational decision making, and research, wherever and whenever these activities take place."

ERIC Clearinghouses collect, abstract, and index education materials for the ERIC database; respond to requests for information in their subject areas; and produce special publications on current research, programs, and practices. The ERIC Processing and Reference Facility is the technical hub of the ERIC system, producing and maintaining the ERIC database and support products for the ERIC system. The ERIC Document Reproduction Service (EDRS) produces and sells paper, microfiche, and electronic copies of documents listed in the ERIC database. ACCESS ERIC coordinates ERIC's outreach, dissemination, and marketing activities; develops systemwide ERIC publications; and provides general reference and referral services.

AskERIC is a personalized, Internet-based service that provides education information to teachers, librarians, counselors, administrators, parents, and others, both nationally and internationally. AskERIC began in 1992 as a project of the ERIC Clearinghouse on Information and Technology at Syracuse University. Today it encompasses the resources of the entire ERIC system and many other sources. Anyone needing the latest information on any education-related topic can simply send a question to askeric@askeric.org. An information specialist will send a personal e-mail response to the question within two working days. Users also receive a list of ERIC citations that deal with their topic, relevant full-text materials (if available), and referrals to organizations and other web-based resources for additional resources.

ERIC is the largest education database in the world, containing more than a million bibliographic records of journal articles, research reports, curriculum and teaching guides, conference papers, and books. Each year approximately 33,000 new records are added. The ERIC database is available in many formats (including printed indexes, CD-ROM, and online) at thousands of locations.

ERIC strives to provide easy and affordable access to education resources for a diverse, global audience. The Internet has been an ideal vehicle for making ERIC's resources available to all interested audiences. To conduct your own tour of ERIC on the Web, start with the ERIC systemwide site at: www.eric.ed.gov.

The section called "ERIC System Directory" contains links to all ERIC-sponsored websites and allows users to send e-mails to ERIC clearinghouses.

Following is a list of ERIC Clearinghouses.

ERIC Clearinghouse on Adult, Career, and Vocational
    Education (ERIC/CE)
E-mail: ericacve@posstbox.acs.ohio-state.edu
URL: http://ericave.org

ERIC Clearinghouse on Assessment and Evaluation (ERIC/TM)
E-mail: ericae@ericae.net
URL: http://ericae.net

ERIC Clearinghouse on Community Colleges (ERIC/JC)
E-mail: ericcc@ucla.edu
URL: http://www.gseis.ucla.edu/ERIC/eric.html

ERIC Clearinghouse on Counseling and Student Services
  (ERIC/CG)
E-mail: ericcass@uncg.edu
URL: http://www.ericcass.uncg.edu

ERIC Clearinghouse on Disabilities and Gifted Education
  (ERIC/EC)
E-mail: ericec@cec.sped.org
URL: http://ericec.org

ERIC Clearinghouse on Educational Management (ERIC/EA)
E-mail: eric@eric.uoregon.edu
URL: http://eric.uoregon.edu

ERIC Clearinghouse on Elementary and Early Childhood
  Education (ERIC/PS)
E-mail: ericeece@uiuc.edu
URL: http://ericeece.org

ERIC Clearinghouse on Higher Education (ERIC/HE)
E-mail: eric-he@eric-he.edu
URL: http://www.eriche.org

ERIC Clearinghouse on Information and Technology (ERIC/IR)
E-mail: eric@ericir.syr.edu or askeric@askeric.org
URL: ERIC/IT: http://ericit.org
URL: AskERIC: http://www.askeric.org

ERIC Clearinghouse on Languages and Linguistics (ERIC/FL)
E-mail: eric@cal.org
URL: http://www.cal.org/ericll

ERIC Clearinghouse on Reading, English, and Communication
(ERIC/CS)
E-mail: ericcs@indiana.edu
URL: http://www.eric.indiana.edu

ERIC Clearinghouse on Rural Education and Small Schools
(ERIC/RC)
E-mail: ericrc@ael.org
URL: http://www.ael.org/eric

ERIC Clearinghouse on Science, Mathematics, and
Environmental Education (ERIC/SE)
E-mail: ericse@osu.edu
URL: http://www.ericse.org

ERIC Clearinghouse on Social Studies/Social Science
Education (ERIC/SO)
E-mail: ericso@indiana.edu
URL: http://www.ericso.indiana.edu

ERIC Clearinghouse on Teaching and Teacher Education
(ERIC/SP)
E-mail: query@aacte.edu
URL: http://www.ericsp.org

ERIC Clearinghouse on Urban Education (ERIC/UD)
E-mail: eric-cue@columbia.edu
URL: http://eric-web.tc.columbia.edu

For a list of the Adjunct Clearinghouses, Affiliate Clearinghouses, Support Components, and ERIC publishers, please visit:
http://www.eric.ed.gov/sites/barak.html

What Is ENC?

by Tracy Crow and Liz Barringer-Smith

ENC's mission is to identify effective curriculum resources, create high-quality professional development materials, and disseminate useful information and products to improve K-12 mathematics and science teaching and learning. ENC:

- Serves all K-12 educators, parents, and students with free products and services.
- Acquires and catalogs mathematics and science curriculum resources, creating the most comprehensive collection in the nation.
- Provides the best selection of math and science resources on the Internet.
- Supports teachers' professional development in math, science, and the effective use of technology.
- Collaborates with the Eisenhower Regional Consortia and many other organizations across the nation to promote education reform.

Located at Ohio State University, ENC is funded through a contract with the U.S. Department of Education's Office of Educational Research and Improvement. To achieve its goals, ENC produces the following products and services.

ENC Online

ENC online (www.enc.org) is the website of the Clearing-house, a gold mine for electronic math and science information. Highlights include:

- The Digital Dozen, a monthly selection of exemplary Internet sites for teachers.
- *Making Schools Work for Every Child*, a collection of materials about achieving educational equity in all schools.
- *ENC Focus*, the online version of ENC's quarterly magazine.

The central feature of ENC Online is the Resource Finder, the electronic catalog of K-12 curriculum and teaching resource materials held in the ENC collection. With input from teachers on the types of information they want to see in a catalog record, ENC developed a unique and comprehensive system. Each record has more than 20 fields, including standard library fields (title, descriptive abstract, author, subjects, publication date, and publisher) as well as fields unique to the needs of educators, including grade level, table of contents, evaluation, and resource type.

As of July 2000, ENC has acquired and catalogued more than 17,000 items, including print materials, video, software, kits, and Internet resources from a variety of sources, including federal and state agencies, commercial publishers, and professional associations.

Free ENC Print Publications

ENC Focus: A Magazine for Classroom Innovators is a free magazine for K-12 classroom teachers, with information of interest to school administrators and policy makers, teacher educators, parents, community members, and all those concerned about education improvement.

Each issues of *ENC Focus* covers an important education theme, with articles on that topic and descriptions of innovative K-12 mathematics and science materials from ENC's vast collection. Other features include essays, classroom stories, and regular columns on such topics as using the Internet in the classroom and applying for grants.

Themes for the magazine include such concepts as Equity, Math and Science in the Real World, and the Standards-Based Classroom. Subscriptions to the print version are free on request; access the online version at www.enc.org.

ENC's Guidebook of Federal Resources for K-12 Mathematics and Science highlights programs for math and science education sponsored and administered by 16 federal agencies and departments. The book is divided into two sections. The first describes the national programs and initiatives of each department or agency; and the second, organized by Eisenhower Consortia region (read about the consortia below) and then divided into state sections, details specific programs available to state and local residents. Each consortium and its activities are described in full, and contact information is included for every program listed.

To request any of ENC's publications, contact ENC at the telephone numbers listed below or visit www.enc.org/contact.

ENC Professional Development Materials

ENC has recently produced its first project specifically for professional developers: *Teacher Change: Improving K-12 Mathematics*. The purpose of this set of materials is to make tools for facilitating discussion and reflection on TIMSS and improving K-12 mathematics available to professional development practitioners. The contents of the package include a set of professional development activities; a variety of teacher narratives and discussion cases; TIMSS data and information, including all released TIMSS test items; and an overview of teacher change by Michael Fullan and Andy Hargreaves.

The contents of this site are online and on CD-ROM. The CD includes a Power Point reader and a Web browser to facilitate its use. Visit www.enc.org/topics/change to use this valuable new tool.

Regional Consortia, Demonstration Sites, and Access Centers

The Eisenhower Regional Consortia and the Clearinghouse work together as a national network to promote reform in math and science. The Consortia have three major goals:

- To identify and disseminate exemplary math and science instructional materials;
- To provide technical assistance to educators in implementing teaching methods and assessment tools; and
- To collaborate with state, local, regional, and national organizations engaged in education improvement.

In cooperation with the consortia, 12 Demonstration Sites — including one at ENC in Columbus, Ohio, and one at George Washington University in Washington, D.C. — were established to provide users with the opportunity to access ENC services electronically.

An ENC Access Center is a facility where the local education community can learn about and access K-12 math and science curriculum resources and the assistance available from ENC and the consortia. There are already more than 160 access centers, with new additions every month.

At the ENC office in Columbus, Ohio, visitors may see items from the national collection. Through the ENC Resource Center, educators can use ENC Online to search for materials and then use the materials on site. A smaller collection of materials, the Capital Collection, is housed at the demonstration site in Washington, D.C.

The Resource Center is staffed to answer reference questions about math and science education. This service is available to all users through e-mail, fax, and telephone. The Technical Help Desk assists ENC Online users with technical questions and problems.

Upcoming Innovations

ENC staff will soon complete an extensive revision of ENC Online, with many improvements. The new design was conceived with the idea that educators need to find good resources quickly. The homepage will serve almost as a site map to the complex set of resources ENC provides. All of the materials most important in an educator's busy life will be highlighted up front: curriculum resources, professional development, funding opportunities, lesson plans, activities, and more. The themes for ENC Focus also will be featured on the homepage, with materials added to those theme sections as they are available.

Contact ENC

Eisenhower National Clearinghouse for Mathematics and
    Science Education
Ohio State University
1929 Kenny Road
Columbus, OH 43210-1079
Toll-free: 800-621-5785
Telephone: (614) 292-7784
Fax: (614) 292-2066
E-mail: info@enc.org
Web: www.enc.org

ENC Reference Staff
Telephone: (614) 292-9734
Toll-free: 800-621-5785
E-mail: library@enc.org

Environmental Education Resources for Multidisciplinary Learning Through Nonformal Education

by Sabiha S. Daudi and Joe E. Heimlich

The learning process through environmental education often has focused on engaging learners through exploration, discovery, investigation, gaming, and hands-on activities. These efforts both draw the learner into the learning exchange and also become the basis for the learning that is to follow. The dominant belief is that the experience does not teach, but prepares the learner for understanding the outcomes of the experience.

Keeping learner engagement through environmental education in mind, we have explored a number of projects that focus on providing an interdisciplinary approach to teaching specific disciplines required to achieve proficiency at different levels of learning for grades K-12. Some of these projects are identified below and illustrate the subject areas addressed by each effort.

Project WILD is an interdisciplinary, supplementary, conservation and environmental education program emphasizing wildlife. The focus is on teachers of grades K-12. The activity guide is divided into seven sections that focus on learners' awareness and appreciation; diversity of wildlife values; ecological principles;

management and conservation; people, culture, and wildlife; trends, issues, and consequences; and responsible human action. Activities also are divided by subject areas to support learning in specific disciplines. These areas include anthropology, art, biology, business education, career education, chemistry, civics, communications, composition, debate, drama, Earth science, economics, English, environmental problems, geography, geometry, government, health, history, home economics, human relations, language arts, life science, mathematics, music, philosophy, photography, psychology, reading, science, social studies, sociology, speech, vocational agriculture, world history, and world geography.

Source: *Project WILD Activity Guide*. Houston, Texas: Council for Environmental Education, 2000. Contact: Project WILD, 5555 Morningside Drive, Suite 212, Houston, TX 77005. Phone: (713) 520-1936; fax: (713) 520-8008. Website: www.project wild.org.

Project Learning Tree (PLT) uses the forest as a "window on the world" to increase students' understanding of our complex environment, to stimulate critical and creative thinking, to develop the ability to make informed decisions, and to instill the confidence and commitment to take responsible action on behalf of the environment. The conceptual framework of PLT's pre-K-8 activity guide focuses on five themes: diversity, interrelationships, systems, structure and scale, and patterns of change. Ninety-six activities in the activity guide enhance learning through the disciplines of visual arts, language arts, mathematics, physical education, science, social studies, and performing arts. Each activity provides comprehensive details for appropriate grade level, subjects, concepts, skills, objectives, materials needed, and time required to complete the activity. The assessments guide the educator toward assessing students' understanding of the concepts covered in the activity and provide opportunities for students to apply the knowledge they have gained.

Source: *Project Learning Tree Environmental Education Pre-K-8 Activity Guide*. Washington, D.C.: American Forest Foundation, 1993. Contact: The American Forest Foundation, 1111 19th Street, N.W., Washington, DC 20036. Website: www.plt.org.

Project WET: Water Education for Teachers aims to facilitate and promote awareness, appreciation, knowledge, and stewardship of water resources through the development and dissemination of classroom-ready teaching aids and through the establishment of state and internationally sponsored Project WET programs. The *Project WET Curriculum and Activity Guide*, for kindergarten through 12th grades, is a collection of innovative, water-related activities that are hands-on, easy to use, and fun. Project WET activities incorporate a variety of formats, such as large and small group learning, whole-body activities, laboratory investigations, discussion of local and global topics, and involvement in community service projects. The activity guide consists of five modules on Watershed, Wetlands, Water/Environmental History, Ground Water, and Water Conservation. The curriculum framework consists of three major areas: conceptual, affective, and skills. The framework is based on current education research, water-related curricula, and national education reforms efforts. It incorporates key concepts related to learning about water and water resources.

Source: *Project WET Curriculum and Activity Guide*. 1995. Bozeman, Mont.: The Watercourse and Council for Environmental Education, 1995. Contact: 201 Culbertson Hall, Montana State University, Bozeman, MT 59717-0570. Phone: (406) 994-5392; fax: (406) 994-1919. E-mail: rwwet@msu.oscs.montana.edu.

Project WET in the City is an urban environmental education program of the Council for Environmental Education (CEE) that focuses on water resources. The program provides an opportunity for young people to participate in engaging, hands-on activities that creatively explore the science of water, its cultural context, and the complex issues surrounding its management and stewardship. The focus on urban issues is a result of the fact that, by the year 2000, more than 90% of the population will live in urban areas. Urban youth seem increasingly disconnected from outdoor pursuits and the natural world. The activities are designed to satisfy the goals of education programs by complementing existing curricula, rather than displacing or adding more concepts. The activities provide many opportunities to address curricular objectives and education standards. Several tools are available to locate activities within the resource guide to meet teachers' needs.

Source: *Project WET in the City Curriculum and Activity Guide.* Houston, Texas: The Watercourse and Council for Environmental Education, 1999. Contact: WET in the City, Council for Environmental Education, 5555 Morningside Drive, Suite 212, Houston, TX 77005-3216. Phone: (713) 520-1936; fax: (713) 520-8008. E-mail: info@wetcity.org. Website: www.wetcity.org.

Windows on the Wild, or WOW, is the World Wildlife Fund's (WWF) environmental education program. The goal of WOW is to educate people of all ages about biodiversity and to stimulate critical thinking, discussion, and informed decision making on behalf of the environment. The program also promotes creative partnerships and interdisciplinary education at all levels. Taking advantage of WWF's unique expertise in addressing biodiversity issues, WOW incorporates current data from projects around the world and draws on the work of many science, development, education, and conservation organizations; government agencies; businesses; and individuals who work closely with WWF to maintain and enhance the Earth's biodiversity. Available are a primer, videos, posters, the Biodiversity Basics teachers guide, student handout books, and supplemental modules.

Source: *Windows on the Wild*. Washington, D.C.: World Wildlife Fund, Acorn Naturalists Publishers, 1999. Contact: World Wildlife Fund, 1250 24th Street, N.W., Washington, DC 20037. Website: www.worldwildlife.org.

North American Journals and Periodicals for Environmental Educators

by Mark A. Miller, Sabiha S. Daudi, and Joe E. Heimlich

The environmental education movement in North America has evolved from conservation education and outdoor education. This transformation has actively progressed in past decades, resulting in the development of activities, teacher training programs, curricula, and teaching and learning resources. Many of these educational resources are readily available to users through the World Wide Web. We have identified some of the periodicals and journals currently available on the database of Educational Resources Information Center/Clearinghouse for Science, Mathematics and Environmental Education (ERIC/CSMEE). We also have highlighted the focus and availability of these journals. These resources can be accessed online at www.eelink.net.

Page down to Class Resources directories, then to EE-Related Education Sites. That will lead you to ERIC. You then will be able to search ERIC databases by following the appropriate prompts. Some issues of the journals and periodicals identified below currently are available through ERIC. Information about subscription and contributions also is provided.

Clearing is a quarterly publication that shares teaching ideas, activities, resources, and a means of communication for formal and nonformal environmental educators for all grade levels and age groups in the United States and Canada. The content primarily reflects activities within the Pacific Northwest — including Oregon, Washington, Idaho, Alaska, and British Columbia — but it is useful and relevant throughout North America. Each issue contains articles on successful classroom programs, perspective articles reflecting on the theory and practice of environmental education, innovative teaching ideas and strategies from educators in the field, a selection of K-12 activities grouped by subject area and grade level, news of environmental education events and programs throughout the Pacific Northwest, and bi-monthly updates written by the current presidents of local, state, and regional environmental education organizations.

Clearing is an independent nonprofit organization, but it is supported in part by the John Inskeep Environmental Learning Center at Clackamas Community College in Oregon City, Oregon. Contact: Clearing, P.O. Box 1432, Oregon City, OR 97045. Phone: (503) 657-6958, ext. 2638; fax: (503) 650-6669. E-mail: clearing@teleport.com. website: www.teleport.com/~clearing.

Green Teacher is a nonprofit quarterly magazine intended to help teachers, parents, and other educators promote environmental and global awareness among young people from kindergarten through senior high school. It offers perspectives on the role of education in creating a sustainable future, practical cross-curricular activities, and reviews of the latest teaching resources. Articles range from such perspective pieces as the role of environmental education or global education in the curriculum to practical classroom-ready activities. Most articles are written by teachers or other educators working in the field of environmental/global education, but several pieces by freelance writers have been published.

Canadian Contact: Green Teacher, 95 Robert Street, Toronto, Ontario M5S 2K5, Canada. Phone: (416) 960-1244; fax: (416) 925-3474.

U.S. Contact: Green Teacher, P.O. Box 1431, Lewiston, NY 14902. E-mail address: greentea@web.net, Website: www. greenteacher.com.

Alternatives is an independent, refereed journal published quarterly by the Environmental Studies Association of Canada (ESAC). The mandate of this journal is to provide critical and informed analyses of environmental issues; to promote an understanding of "environment" in the broadest sense of the word, including social and political dimensions of environments; to reflect a Canadian perspective informed by an understanding of global issues; to stimulate dialogue and exchange of information among environmental activists, academics, and professionals; and to create a publishing opportunity for Canadian scholars and professionals.

Contact: Environmental Studies, University of Waterloo, Waterloo, N2L 3G1, Canada. Phone: (519) 888-4567, ext. 6783; fax: (519) 746-0292. E-mail: alternat@fes.uwaterloo.ca. Website: www.fes.uwaterloo.ca/alternatives.

Earthwatch is a journal published by the Earthwatch Institute. Earthwatch Institute is an international nonprofit organization that supports scientific field research worldwide to improve our understanding and management of the Earth. The institute's mission is to promote sustainable conservation of our natural resources and cultural heritage by creating partnerships between scientists, educators, and the general public. Through the participation of volunteers in field research, Earthwatch helps scientists gather vital data that empower individuals and governments to act wisely as global citizens.

Contact: Earthwatch Institute Headquarters, 680 Mt. Auburn St., Box 9104, Watertown, MA 02471-9104. Phone: 800-776-0188; fax (617) 926-8532. E-mail: info@earthwatch.org. Website: www.earthwatch.org.

The Environmentalist is a quarterly international journal that acts as a catalyst for environmental education, identifying available education opportunities and providing necessary guidelines and frameworks for defining the more viable management mechanisms useful to industry, government policy makers, and environmental professionals. The *Environmentalist* publishes the critical but constructive views from both industrialists and ecologists, through challenging guest editorials, in-depth articles, interviews, and the news and comments columns. It contains elements applicable to the education and training of people at various levels, be it formal or nonformal schooling, specialist training, retaining of decision makers, or communication to the public at large.

Contact for North and South America: The Environmentalist, P.O. Box 358, Accord Station, Hingham, MA 02018-0358. Phone: (781) 871-6600; fax: (781) 681-9045. E-mail: kluwer@wkap.com.

Contact for outside the Americas: The Environmentalist, P.O. Box 989, 3300 AZ Dordrecht, The Netherlands. Phone: (+31) 78 639 23 92; fax: (+31) 78 639 22 54. E-mail: services@wkap.nl. Website: www.wkap.nl.

ERIC journals (citations identified by an EJ number) are available in your local library or via interlibrary loan services, from the originating journal publisher, or for a fee from the following article reproduction vendors: CARL UnCover S.O.S.; e-mail sos@carl.org. Phone (800) 787-7979. ISI Document Solution: Phone: (800) 336-4474 or (215) 386-4399; e-mail: ids@isinet.com; or online order form: www.isidoc.com.

A Plethora of Learning Resources for Environmental Education in the U.S.: Part I

by Mark A. Miller, Sabiha S. Daudi, and Joe E. Heimlich

The environmental education movement has come a long way since it inception in the 1970s. This is particularly evident in the United States, where the movement is well supported by development of various teaching and learning resources. Human resource development has been highlighted by various training programs available for teachers and nonformal educators. These efforts also have been strengthened by the active involvement of government and public and private sectors, who focused on developing information resources and publishing journals and periodicals that deal with current environmental issues and concerns.

Some of these resources are indexed and abstracted by Educational Resources Information Center/Center for Science, Mathematics, and Environmental Education (ERIC/CSMEE). We have identified some of the periodicals and journals currently available on the database of ERIC/CSMEE. These resources can be accessed online at www.eelink.net.

Page down to Class Resources directories, then to EE-Related Education Sites, which will lead you to ERIC. You then will be able to search ERIC databases by following the appropriate prompts. Some issues of the journals and periodicals identified below currently are available through ERIC. Information about subscriptions also is provided.

American Forests is a quarterly magazine published by the organization of the same name to help people plant and care for trees for ecosystem restoration and healthier communities. American Forests is the nation's oldest citizen conservation organization. Since 1875 it has worked to ensure a sustainable future for the nation's forests.

Contact: American Forests, P.O. Box 2000, Washington, DC 20013. Phone: 800-368-5748 or (202) 955-4500. E-mail : (membership questions or requests) member@amfor.org. Website: www.americanforests.org.

The Amicus Journal is a quarterly publication of the Natural Resources Defense Council (NRDC). NRDC is a national nonprofit organization dedicated to protecting the world's natural resources and ensuring a safe and healthy environment for all people. The journal includes thought and opinion for the general public on environmental affairs, particularly those relating to policies of national and international significance. In keeping with NRDC's program of public-interest advocacy and its efforts to educate the public about environmental protection, this magazine offers essays, news analyses, features, profiles, reviews, and poetry. It also integrates news about NRDC, its staff, and their work with news about the environmental movement as a whole.

Contact: NRDC, 40 West 20th Street, New York, NY 10011. Phone: (212) 727-2700. E-mail : amicus@nrdc.org. Website: www.nrdc.org.

Environment is published 10 times per year by Heldref Publications, a division of the nonprofit Helen Dwight Reid Educational Foundation. *Environment* provides authoritative yet readable analyses of key environmental science and policy issues. Readers enjoy the magazine's comprehensive articles written by the world's foremost scientists and policy makers, detailed reviews of major governmental and institutional reports, pointers to the best environmental websites and other digital media, book recommendations, commentaries that expand on controversial topics, and news briefs.

Contact: Environment, Heldref Publications, 1319 18th Street, N.W., Washington, DC 20036-1802. Phone: (202) 296-6267. Subscription orders: 800-365-9753; fax (202) 293-6130. Website: www.heldref.org.

Focus journal and other selections are published quarterly by the Carrying Capacity Network (CCN). "Carrying Capacity Network confronts the controversial issues, makes the difficult choices, and adopts the innovative approaches necessary to meet the challenges facing our nation. CCN's action-oriented initiatives focus on achieving national revitalization, population stabilization, immigration reduction, economic sustainability, and resource conservation. With the Network Bulletin, Focus, Carrying Capacity Research Issues Series, and our Fax Alert System, CCN offers its members the information and tools necessary to facilitate effective action."

Contact: CCN, 2000 P Street Suite 240, N.W., Washington, DC 20036-5915. Phone: (202) 296-4548; fax (202) 296-4609. E-mail: ccn@us.net. Website: www.carryingcapacity.org.

Journal of Environmental Health is published 10 times per year by the National Environmental Health Association (NEHA). Their mission is to advance the environmental health and protection profession for the purpose of providing a healthful environment for all. Advancement has been defined by NEHA in terms of both education and motivation. The basis for the association's activities is the belief that the professional who is educated and motivated is the professional who will make the greatest contribution to the goals of a healthful environment.

Contact: National Environmental Health Association, 720 S. Colorado Boulevard, South Tower, Suite 970, Denver, CO 80246-1925. Phone: (303) 756-9090; fax (303) 691-9490. E-mail: staff@neha.org. Website: www.neha.org.

Journal of Natural Resources and Life Sciences Education is a yearly international journal published by the American Society of Agronomy (ASA). Cooperating organizations include the American Association for Agricultural Education, the American Institute of Biological Sciences, the American Phytopathological Society, the American Society for Horticultural Science, Ecological Society of America, the Crop Science Society of America, the Entomological Society of America, and the Soil Science Society of America. Because of their common interests, the American Society of Agronomy (ASA), Crop Science Society of America (CSSA), and Soil Science Society of America (SSSA) share a close working relationship, as well as the same headquarters office staff. However, each of the three societies is autonomous, has its own bylaws, and is governed by its own board of directors. The societies' goals are: To promote effective research and teaching, to foster high educational standards, to disseminate agronomic sciences information, to encourage professional growth, and to interact with organizations sharing similar goals. Society members are dedicated to the conservation and wise use of natural resources to produce food, feed, and fiber crops while maintaining or improving the environment.

Contact: American Society of Agronomy, 677 S. Segoe Rd., Madison, WI 53711. Phone: (608) 273-8080; fax: (608) 273-2021. E-mail: ipopkewitz@agronomy.org. Website: www.jnrlse.org.

A Plethora of Learning Resources for Environmental Education in the U.S.: Part II

by Mark A. Miller, Sabiha S. Daudi, and Joe E. Heimlich

Continuing on our quest for teaching and learning resources for environmental education that are available in the United States, it is interesting to note that there are more publications that highlight environmental concerns and related issues. These resources take into consideration humanistic views, as well as scientific and ecological dimensions. They have supported a need for information that was expressed by academia and communities alike.

Some of these resources are indexed and abstracted by Educational Resources Information Center/Center for Science, Mathematics, and Environmental Education (ERIC/CSMEE). We have identified some of the periodicals and journals currently available on the database of ERIC/CSMEE. These resources can be accessed at www.eelink.net.

Page down to Class Resources directories, then to EE-Related Education Sites, which will lead you to ERIC. You then will be able to search ERIC databases by following the appropriate prompts. Some issues of the journals and periodicals identified below currently are available through ERIC. Information about subscriptions also is provided.

The Ecologist is published 10 times per year by Ecosystems Ltd. The Ecologist states that "[they] consider that every one of the depressing problems which [they] have described ad nauseam in The Ecologist during the past 29 years is the inevitable consequence of the policies imposed on society by our political and industrial leaders. If this is so, then the only responsible and indeed the only honest thing to do is to stand back and reconsider the basic assumptions on which these policies are based and the basic assumptions that unfortunately underlie the disciplines (as they are taught today) into which modern knowledge is divided."

Contact the Subscription Office, Cissbury House, Furze View, Five Oaks Road, Slinfold, West Sussex RH13 7RH, United Kingdom.

U.S. Office: 1920 Martin Luther King Jr. Way, Berkeley, CA 94704, USA. Phone: (510) 548-2032; fax: (510) 548-4916. E-mail: sgc@mag-subs.demon.co.uk. Website: www.theecologist. org.

National Parks is the bimonthly magazine of the National Parks and Conservation Association. NPCA's mission, to protect and enhance America's National Park System for present and future generations, can be achieved only through a well-educated and committed park constituency. Public education through publications and public outreach to inspire activism is key to their efforts. The activism of members and friends backs NPCA's advocacy role for the parks among lawmakers and park managers, calling for the strongest protections for our national treasures. When laws protecting parks are not followed, they use litigation as a way to further ensure that our parks are protected.

Contact: National Parks, 1776 Massachusetts Ave., NW, Washington, DC 20036. Phone: (202) 223-6722 or 800-NAT-PARK; fax: (202) 659-8178. E-mail: npmag@npca.org. Website: www.npca.org.

Nature Conservancy is published bimonthly by the Nature Conservancy. Their mission is to preserve plants, animals, and natural communities that represent the diversity of life on Earth by protecting the lands and waters they need to survive. The Nature Conservancy operates the largest private system of nature sanctuaries in the world — more than 1,500 preserves in the United States alone. Some are very small, others cover thousands of acres. All of them safeguard imperiled species of plants and animals.

Contact: The Nature Conservancy, 4245 North Fairfax Drive, Suite 100, Arlington, VA 22203-1606. Phone: (703) 841-5300; fax: (703) 841-4100. Website: www.tnc.org.

Our Planet is a bimonthly magazine of the United Nations Environment Programme (UNEP). UNEP's mission is to provide leadership and encourage partnership in caring for the environment by inspiring, informing, and enabling nations and peoples to improve their quality of life without compromising that of future generations.

Contact: Our Planet, IWSS Ltd, P.O. Box 119, Stevenage, Hertfordshire, SG1 4TP, UK. Website: www.unep.org.

Population Research and Policy Review is a bimonthly publication from Kluwer Academic Publishers. It seeks to publish quality material of interest to professionals working in the field of population and those fields that intersect and overlap with population studies. The publication includes demographic, economic, social, political, and health research papers and related contributions that are based on either the direct scientific evaluation of particular policies or programs or general contributions intended to advance knowledge that informs policy and program development.

Contact: Population Research and Policy Review, P.O. Box 358, Accord Station, Hingham, MA 02018-0358. Phone: (781) 871-6600; fax: (781) 681-9045. E-mail: kluwer@wkap.com. Website: www.wkap.nl.

Streetwise is a quarterly publication from Places for People, the National Association for Urban Studies. *Streetwise* aims to inform and inspire people interested in urban environmental education and the process of public participation in positive change. Places for People is concerned with the processes of change in the urban environment. They encourage effective participation in the creation of sustainable communities and provide a forum for people with an interest in the built environment and urban environmental education. Its objectives are to promote education and action for a sustainable urban environment in the framework of Agenda 21; to support the right of everyone to have a voice in the future of the places in which they live, work, or play; and to encourage practical cooperation in improving the urban environment.

Contact: Streetwise, c/o ETP, 9 South Road, Brighton BN1 6SB, United Kingdom. Phone and fax: 01273-542660. E-mail: streetwise@pobox.com. Website: pobox.com/~streetwise.

ERIC Journals (citations identified by an EJ number) are available in your local library or through interlibrary loan services, from the originating journal publisher, or for a fee from the following article reproduction vendors: CARL UnCover S.O.S. E-mail: sos@carl.org. Phone: 800-787-7979. ISI Document Solution. E-mail: ids@isinet.com. Phone: 800-336-4474, (215) 386-4399, or use online order form: www.isidoc.com.

A Plethora of Learning Resources for Environmental Education in the U.S.: Part III

by Mark A. Miller, Sabiha S. Daudi, and Joe E. Heimlich

A collage of learning and teaching resources is available to environmental educators to enhance their understanding of global and regional environmental issues. In order to support development of effective teaching plans and projects about the natural environment and interactions between nature and living things, a number of resources currently are available on the database of Educational Resources Information Center/Center for Science, Mathematics, and Environmental Education (ERIC/CSMEE). We have identified some of the periodicals and journals currently available on the database of ERIC/CSMEE. These resources can be accessed at www.eelink.net.

Page down to Class Resources directories, then to EE-Related Education Sites, which will lead you to ERIC. You then will be able to search ERIC databases by following the appropriate prompts. Some issues of the journals and periodicals identified below currently are available through ERIC. Information about subscriptions also is provided.

The Workbook is published quarterly by the Southwest Research and Information Center. The *Workbook* is a fully indexed catalogue of sources of information about environmental, social, and consumer problems. It is aimed at helping people gain access to vital information that can help them assert control over their own lives. The Southwest Research and Information Center (SRIC) exists to provide timely, accurate information to the public on matters that affect the environment, human health, and communities in order to protect natural resources, promote citizen participation, and ensure environmental and social justice now and for future generations.

Contact: The Workbook, 105 Stanford Drive S.E., P.O. Box 4524, Albuquerque, NM 87196. Phone: (505) 346-1455; fax: (505) 346-1459. E-mail: WorkbookEd@aol.com. Website: www.SRIC.org

World Watch is a bimonthly, not-for-profit magazine that tracks key indicators of the Earth's well-being. They monitor and evaluate changes in climate, forest cover, population, food production, water resources, biological diversity, and other key trends and identify and analyze the most effective strategies for achieving a sustainable society — including those that come from the advances of science and technology, the rethinking of traditional economics, and the neglected wisdom of now-vanishing indigenous peoples. Their goal is to provide policy makers, educators, researchers, reporters, and concerned individuals worldwide with the information they need to make decisions that will lead to a sustainable economy.

Contact: World Watch, 1776 Massachusetts Ave., NW, Washington, DC 20036. Phone: (202) 452-1999; fax: (202) 296-7365. E-mail: worldwatch@worldwatch.org. Website: www. worldwatch.org

W.O.W. Wild Outdoor World is published bimonthly by the Rocky Mountain Elk Foundation. It aims to enhance future generations' awareness of and appreciation for the natural world. The magazine helps foster a sense of the need for conservation through articles focused on wild animals and their habitats, wildlife management, and healthy outdoor activity. Fourth-graders

throughout the U.S. receive the magazine free through a distribution program sponsored by state and federal natural resource
agencies, private groups, and interested individuals.

Contact: Wild Outdoor World, P.O. Box 8249, Missoula, MT
59807-8249. Phone: 888-301-5437, ext. 401.

Environmental Education, published by the National Association for Environmental Education (UK), is issued free to members three times yearly. The National Association for
Environmental Education (UK) states that environmental education involves the recognition of values, clarification of concepts,
and development of the skills and attitudes needed to understand
the inter-relatedness of man, his culture, and his biophysical
surroundings; practice in decision making; concern for environmental quality; and the adoption of a code of behavior. NAEE is
the association of teachers, lecturers, and others concerned with
education and the environment. Its members work in all types of
schools, colleges, and universities.

Contact: NAEE Office, Wolverhampton University, Walsall
Campus, Gorway Rd., Walsall, West Midlands WS1 3BD, United
Kingdom.

Journal of Interpretation Research is a journal published in association with the National Association for Interpretation. The pur-
poses of the journal are to communicate empirical research dealing
with interpretation and to provide a forum for scholarly discourse
about issues facing the profession of interpretation. In recognition
of how difficult it is for interpreters to keep up with the growing and
diverse body of relevant literature, the journal publishes reviews of
recent books, professional meetings and workshops, government
publications, and original literature reviews and bibliographies dealing with interpretation. The *Journal of Interpretation Research*
takes a broad view of the field of interpretation and publishes man-
uscripts from a wide range of academic disciplines. The journal also
is published in Spanish as *Investigaciones en Interpretacion.*

Contact: National Association for Interpretation, P.O. Box
2246, Fort Collins, CO 80522. Phone: 888-900-8283, (970) 484-

8283; fax (970) 484-8179. E-mail: naicom@aol.com. Website: www.interpnet.com or www.interpnet.org.

Pennsylvania Alliance for Environmental Education Journal is published quarterly by the Pennsylvania Alliance for Environmental Education (PAEE). The mission of PAEE is to promote environmental education activities and efforts throughout Pennsylvania. The ultimate outcome of such a mission is a citizenry that understands and appreciates its relationship to the natural world, that recognizes and accepts responsibility for its effect on natural systems, and that is motivated to take positive action to solve environmental problems.

Contact: PAEE, 225 Pine St., Harrisburg, PA 17101. Phone: (717) 236-3599 or (717) 238-0426; fax: (717) 238-2436. Website: www.telesync.com/paee.

Environmental Education Resources for Teaching Various Disciplines: Part I

by Mark A. Miller, Sabiha S. Daudi, and Joe E. Heimlich

Environmental education has been identified as a source of inter-disciplinary and transdisciplinary educational activities for teaching and learning different core curricular subjects. It aims to provide a convergence of teaching and learning methods to K-12 teachers and nonformal educators. Environmental education resources are available through the databases of Educational Resources Information Center/Clearinghouse for Science, Mathematics, and Environmental Education (ERIC/CSMEE) and the Eisenhower National Clearinghouse (ENC). These resources can be accessed at www.eelink.net.

Page down to Class Resources directories, then to EE-Related Education Sites. That will lead you to ERIC or ENC. You then will be able to search ERIC and ENC databases by following the appropriate prompts.

A number of resources also are available through journals that have environmental education as one of their areas of interest. ERIC journals (citations identified by an EJ number) are available in your local library or by interlibrary loan, from the originating journal publisher, or for a fee from the following article reproduction vendors: CARL UnCover S.O.S. E-mail: sos@carl.org. Phone: (800) 787-7979. ISD Document Solution. E-mail:

ids@isinet.com; phone: 800-336-4474 or (215) 386-4399; or online at www.isidoc.com.

We have identified some journals that have a focus on different disciplines of education, as well as environmental education. Please note that these are only the ones being indexed or abstracted by ERIC and do not constitute a comprehensive list of journals. ERIC reviews journals only in English and French at this time.

Agriculture

Agricultural Research is a monthly publication by the Agricultural Research Service, U.S. Department of Agriculture (USDA). The Agricultural Research Service (ARS) is the principal in-house research agency of the USDA. It is one of the four component agencies of the Research, Education, and Economics (REE) mission area. ARS conducts research to develop and transfer solutions to agricultural problems of high national priority and provides information access and dissemination to ensure high-quality, safe food and other agricultural products; assess the nutritional needs of Americans; sustain a competitive agricultural economy; enhance the natural resource base and the environment; and provide economic opportunities for rural citizens, communities, and society as a whole.

Contact: New Orders, Superintendent of Documents, P.O. Box 371954, Pittsburgh, PA. Phone: (202) 512-1800; fax: (202) 512-2250. Website: www.ars.usda.gov/is/AR.

Journal of Pesticide Reform is published four times a year by the Northwest Coalition for Alternatives to Pesticides (NCAP). NCAP provides assistance in developing model policies to protect our groundwater, food supply, and forest watersheds from pesticide contamination; successful model pest management policies for school grounds, roadsides, national forests, and more; information on hundreds of pesticides and alternatives for many pest problems; updates on citizen reform efforts and policy initiatives from across North America through the journal; direct assistance and referrals for pesticide exposure victims; and organizing assistance for citizens working for policy reform in their communities.

Contact: Northwest Coalition for Alternatives to Pesticides (NCAP), P.O. Box 1393, Eugene, OR 97440. Phone (541) 344-5044; fax (541) 344-6923. Email: info@pesticide.org. Website: www.pesticide.org/default.htm.

Geology

Environmental Geology, International Journal of Geosciences, is a monthly publication by Springer-Verlag. It is an international multidisciplinary journal concerned with all aspects of interactions between humans, ecosystems, and the Earth. Dissemination of knowledge on techniques, methods, approaches, and experiences aims at improvement and remediation of the environment as habitat for life on Earth. In pursuit of these topics, the geoscientific disciplines are invited to contribute their knowledge and experience. Major disciplines include hydrogeology, hydrochemistry, geochemistry, geophysics, engineering geology, soil science, and geomicrobiology.

Contact: Springer-Verlag Berlin, New York Inc., Journal Fulfillment Services, Dept., 333 Meadowlands Parkway, Secaucus, NJ 07094. Phone: (201) 348-4033; fax: (201) 348-4505. Email: subscriptions@springer.de. Website: link.springer.de.

Environmental Monitoring and Assessment is published 18 times per year by Kluwer Academic Publishers. It emphasizes technical developments and data arising from environmental monitoring and assessment; the use of scientific principles in the design of monitoring systems at the local, regional, and global scales; and the use of monitoring data in assessing the consequences of natural resource management actions and pollution risks to man and the environment. The journal covers a wide range of pollutants and examines monitoring systems designed to estimate exposure both at the individual and population levels.

The journal also focuses on the development of monitoring systems related to the management of various renewable natural resources in, for instance, agriculture, fisheries, and forests. The scope of the journal extends to the use of monitoring in pollution assessment; and particular emphasis is given to the synthesis of monitoring data with toxicological, epidemiological, and health data, as well as with pre-market screening results. The journal also includes research and monitoring systems that help assess anthropogenic effects on natural resources and the environment from numerous activities, such as harvesting, development, and land use changes. Geographic information system analyses and remote sensing studies relating such activities to land cover changes that affect, for example, biodiversity and global climate change also are within the purview of the journal.

Contact: Environmental Monitoring and Assessment, P.O. Box 358, Accord Station, Hingham, MA 02018-0358 (North and South America); Environmental Monitoring and Assessment, P.O. Box 989, 3300 AZ Dordrecht, The Netherlands (rest of the world). Phone: (781) 871-6600 (US) or (+31) 78 639 23 92 (rest of the world); fax: (781) 681-9045 (US) or (+31) 78 639 22 54 (rest of the world). E-mail: kluwer@wkap.com (US), services@wkap.nl (rest of the world). Website: www.wkap.nl.

Coast & Sea is published twice a year by Sea Grant, Louisiana, as part of the National Sea Grant College Program maintained by the National Oceanic and Atmospheric Administration of the U.S. Department of Commerce. Sea Grant is a unique partnership with public and private sectors, combining research, education, and technology transfer for public service. Sea Grant is the national network of universities meeting changing environmental and economic needs of people in our coastal, ocean, and Great Lakes regions.

Contact: Coast & Sea, Louisiana Sea Grant Communications Office, Louisiana State University, Baton Rouge, LA 70803. Phone: (225) 388-6448; fax: (225) 388-6331. Website: www.laseagrant.org.

Environmental Education Resources for Teaching Various Disciplines: Part II

by Mark A. Miller, Sabiha S. Daudi, and Joe E. Heimlich

The databases of Educational Resources Information Center/ Clearinghouse for Science, Mathematics and Environmental Education (ERIC/CSMEE) and the Eisenhower National Clearinghouse (ENC) have a multitude of resources available to educators. These resources can be accessed at www.eelink.net.

Page down to Class Resources directories, then to EE-Related Education Sites. That will lead you to ERIC or ENC. You then will be able to search ERIC and ENC databases by following the appropriate prompts.

A number of resources also are available through journals that have environmental education as one of their areas of interest. ERIC journals (citations identified by an EJ number) are available in your local library or by interlibrary loan, from the originating journal publisher, or for a fee from the following article reproduction vendors: CARL UnCover S.O.S. E-mail: sos@carl. org. Phone: (800) 787-7979. ISI Document Solution. E-mail: ids@isinet.com; phone: 800-336-4474 or (215) 386-4399; or online at www.isidoc.com.

We have identified some journals available at the database of Educational Resources Information Center (ERIC) that will help K-12 and nonformal educators reinforce their lessons in science and other disciplines.

Science and Marine Education

Current, The Journal of Marine Education is published up to four times a year for members of the National Marine Educators Association (NMEA). NMEA brings together those interested in the study and enjoyment of the world of water. Affiliated with the National Science Teachers Association, NMEA includes professionals with backgrounds in education, science, business, government, museums, aquariums, and marine research, among others.

Contact: NMEA, P.O. Box 1470, Ocean Springs, MS 39566-1470. Phone: (228) 374-7557. E-mail: johnette.borsage@usm.edu. Website: www.marine_ed.org.

Revue Des Sciences de L'Eau, Journal of Water Science, is published quarterly by Lavoisier. It is a multidisciplinary scientific journal publishing the most recent research in five broad categories: hydrology, hydrogeology, and water resources management; physical chemistry of the aquatic environment; hydrobiology, microbiology, toxicology, and ecotoxicology; quality and treatment of drinking water; and wastewater treatment. Subscribers from the Americas (Canada, United States, Mexico, and South American countries) should contact the INRS-Eau. Those from Europe, Africa, and Asia should contact Lavoisier Abonnements.

Contact: Lavoisier Abonnement, 14 rue de Provigny, 94236 Cachan Cedex, France; phone: +33 (1) 47.40.67.00; fax: +33 (1) 47.40.67.02. E-mail: abo@lavoisier.fr. Website: http://www.lavoisier.fr. Or Contact: INRS-Eau, 2800 rue Einstein, Casier Postal 7500, Sainte-Foy (Quebec), G1V 4C7 Canada. Phone: (418) 654-2649; fax: (418) 654-2600. E-mail: rseinfo@inrs-au.uquebec.ca. Website: http://www.inrs-eau.uquebec.ca.

Science and Technology

Environmental Science & Technology is published semi-monthly by the American Chemical Society (ACS). It features peer-reviewed research and news reports concerning all aspects of environmental science, technology, and policy analysis. Coverage also includes interpretive articles by invited experts and analysis of the scientific aspects of environmental management. *Environmental Science & Technology* provides unique, valuable insights into scientific studies of the chemical and biological nature of the environment, environmental changes through pollution or other global changes, and new approaches to waste treatment and minimization.

Contact: American Chemical Society, Dept. L-0011, Columbus, OH 43268-0011. Phone: (614) 447-3776 or 800-333-9511; fax: (614) 447-3671. E-mail: service@acs.org. Website: pubs.acs.org.

Science and Solar Energy

Solar Today is a bimonthly publication of the American Solar Energy Society (ASES). It is an award-winning magazine that covers all solar technologies, from photovoltaics to climate-responsive buildings to wind power. Regular topics include building case studies, energy policy, and community-scale projects. ASES is a national organization dedicated to advancing the use of solar energy for the benefit of U.S. citizens and the global environment. ASES promotes the widespread, near-term, and long-term use of solar energy. It is the United States section of the International Solar Energy Society.

Contact: American Solar Energy Society, 2400 Central Ave., Suite G-1, Boulder, CO 80301. Phone: (303) 443-3130; fax: (303) 443-3212. E-mail: ases@ases.org. Website: www.ases.org.

National Wildlife is published bimonthly by the National Wildlife Federation (NWF). The NWF mission is "to educate, inspire and assist individuals and organizations of diverse cultures to conserve wildlife and other natural resources and to protect the Earth's environment in order to achieve a peaceful, equitable and sustainable future."

Contact: National Wildlife, 8925 Leesburg Pike, Vienna, VA 22184. E-mail: pubs@nwf.org. Website: www.nwf.org.